RAND NATIONAL DEFENSE RESEARCH INSTITUTE

Enhancing Management of the Joint Future Vertical Lift Initiative

Jeffrey A. Drezner, Parisa Roshan, Thomas Whitmore

Prepared for the Office of the Joint Chiefs of Staff

For more information on this publication, visit www.rand.org/t/RR2010

Library of Congress Cataloging-in-Publication Data is available for this publication.
ISBN: 978-0-8330-9837-5

Published by the RAND Corporation, Santa Monica, Calif.

Cover: CV-22B Osprey (U.S. Air Force photo by Airman 1st Class Trevor T. McBride)

Preface

The history of joint acquisition programs in the U.S. Department of Defense (DoD) reveals mixed outcomes—some positive, some negative. Joint program management is intended to reduce the investment, management, administrative, and support costs throughout a program's life cycle and spread risks across participating services. Increased commonality theoretically yields cost savings from reduced duplication of development activities and economies of scale that are realized during the production and support phases. However, joint management introduces significant complexities, while commonality introduces significant technical challenges that can contribute to cost growth, schedule delays, and performance shortfalls, all of which detract from the benefits expected from commonality.

The joint Future Vertical Lift (FVL) initiative asked the RAND Corporation to examine joint management constructs and recommend strategies for improving both its internal organizational structure and its alignment with key external bodies. Based on a review of historical joint initiatives, as well as a review of relevant business management literature, we identify some of the factors affecting joint program outcomes and recommend ways to apply those lessons to the management of FVL. This study should be of interest to senior leaders and program managers in DoD planning or currently involved in joint programs, as well as to the broader DoD acquisition community.

This study was jointly funded by FVL stakeholders and sponsored by the Deputy Director, J-8, Office of the Joint Chiefs of Staff and the FVL initiative. The study was conducted within the Acquisition and Technology Policy Center of the RAND National Defense Research

Institute, a federally funded research and development center sponsored by the Office of the Secretary of Defense, the Joint Staff, the Unified Combatant Commands, the Navy, the Marine Corps, the defense agencies, and the defense Intelligence Community.

For more information on the RAND Acquisition and Technology Policy Center, see www.rand.org/nsrd/about/atp or contact the director (contact information is provided on the web page).

Contents

Figures and Tables

Figures

Tables

Summary

The Future Vertical Lift (FVL) initiative grew out of a May 2008 Secretary of Defense directive to the Office of the Under Secretary of Defense for Acquisition, Technology and Logistics (OUSD[AT&L]) and the Joint Staff to examine a joint approach to the development of FVL aircraft for all military departments. The fiscal year 2009 National Defense Authorization Act also called for this effort.[1] Joint management of major defense acquisition programs (MDAPs) has often been posited as an approach to reduce redundancy and duplication across the services; avoid duplicate investment in technologies and systems; and generate savings in the production, operations, and support phases of a given program. However, joint program management introduces significant complexity in terms of developing requirements and managing and overseeing the acquisition program and its requisite processes (e.g., design, development, manufacturing, test). Additionally, joint program management often increases the risk of technical complications as the interests and potentially conflicting requirements of the different services are traded off and balanced.

In May 2015, the RAND National Defense Research Institute was tasked with examining alternative joint management constructs suitable for FVL. As part of this effort, we were asked to review historical experience with joint program management, identify factors affecting program success, and recommend ways to apply those lessons to the FVL management organization. We conducted discussions

---

[1] Deputy Secretary of Defense, *Future Vertical Lift: A Strategic Plan for United States Department of Defense Vertical Lift Aircraft*, Version 2, Washington, D.C.: U.S. Department of Defense, October 1, 2011.

with FVL stakeholders and participants, as well as representatives from select joint or multiservice programs. Lastly, we reviewed the organizational theory and business management literature for lessons that could be applied in the FVL context.

FVL is envisioned as a family of systems (FoS) with distinct programs based on different capability sets and performance specifications. Each individual program would include participation of the stakeholders whose capability needs are met by a specific capability set. The FVL initiative's current management construct designates the Army as the lead agency and OUSD(AT&L) as the milestone decision authority. A 2-star/SES–level executive steering group (ESG) was established as the policymaking and decisionmaking body for FVL. A Joint Council of Colonels (JCoC) is the action arm of the ESG and was charged with execution and implementation of the *FVL Strategic Plan*. The JCoC consists of colonel-level/O-6 representatives for each of the services, U.S. Special Operations Command, the Joint Staff, and the Office of the Secretary of Defense staff. A Joint Coordination and Integration Cell (JCIC) is responsible for coordination, communication, and synchronization throughout the FVL community. The ESG also chartered four Integrated Product Teams (IPTs): acquisition, requirements, science and technology, and common systems. Neither the ESG nor the JCoC manages requirements or acquisition processes; they are not U.S. Code (U.S.C.) Title 10 bodies.[2] Rather, they make recommendations to the services. Because the members of the ESG represent the service requirements and acquisition communities, the appropriate 10 U.S.C. service office or agency is represented. The current management structure thus facilitates improved communication and information-sharing across the services and acts as a forum for both formal and informal joint decisionmaking regarding FVL.

There is a wide range of management constructs possible for FVL. The specific advantages and disadvantages—benefits and costs—of each management construct are contingent on a clear definition of exactly

---

[2] Title 10 of the U.S. Code (10 U.S.C.) is the law giving the military departments the authority and responsibility to train and equip the armed forces. OUSD(AT&L) also has 10 U.S.C. authority through delegation by the Secretary of Defense.

what is being acquired by whom and the business, organizational, and the budgetary environment. That is, the advantages and disadvantages of any particular management construct are context-dependent.

The study's key findings include the following:

- Joint management and commonality are not the same. Joint management is an organizational construct established to jointly manage a program with multiple participants. In contrast, commonality is a technical design concept in which different systems use the same components or subsystems. In general, it is commonality (in some form) that is responsible for cost savings or other benefits that are attributed to joint management. Commonality may emerge organically (from the bottom up) and be managed successfully without a formal joint management construct.
- Voluntary participation is a key enabler of successful joint initiatives. This ensures that participants are self-motivated and self-organizing.
- Requirements must be substantially the same across all stakeholders. This means that the basic technical characteristics of the system (or subsystems and components) must be similar enough that compromises to accommodate diverse requirements do not impose additional costs (in dollars, risk, or performance) on the system. This is extremely challenging for a complex multiservice weapon system program, but it may be somewhat more manageable at the subsystem or component level.
- Any formal joint or multiservice management construct should be founded on comprehensive planning, which includes delineating the roles, responsibilities, authorities or participants, decision and oversight processes, and other rules of engagement. These should be codified in a memorandum of agreement.

We believe that even if FVL becomes a series of independent single-service programs, some degree of commonality will emerge organically. If FVL stakeholders want to increase commonality above this level, then some form of higher-level coordinating body or active management construct is required. This construct does not need to be centered on joint management, but some degree of multiservice participation is required. An organization to facilitate and oversee information flow, science and technology investment and outputs, requirements development, and budget planning would be needed. Continuation of the ESG, JCoC, IPTs, and JCIC is likely the best option for this, although the frequency of meetings and activities can be adjusted to meet the needs of the participants in different phases of planning, analysis, and program execution.

There is no evidence that joint management produces net benefits for a complex program, let alone a series of complex programs as envisioned in the FVL FoS. Joint programs seen as successful tend to be relatively less complex (i.e., Joint Light Tactical Vehicle; Joint Direct Attack Munition), whereas more-complex programs do not demonstrate the expected benefits and, in some cases, may incorporate significantly less commonality than originally expected (i.e., F-35 Joint Strike Fighter). Therefore, we recommend that FVL not implement a purely joint construct at this time. A lead service (executive agent) construct appears to be the most likely to be successful at the system level. A lead service approach can include direct multiservice participation (i.e., representation for other services in key leadership positions within the program office) or more of a leader-follower model of derivative commonality, in which the follower services use the basic system (i.e., airframe and engine) but incorporate their own unique mission systems separately.

However, FVL does represent an opportunity to more fully explore commonality and constructs for managing commonality, analyze the key actions or decisions that generate commonality, and better understand the costs and benefits associated with commonality. A thorough technical analysis of whether commonality makes sense, independent of management constructs, should be conducted first. The analysis should include whether commonality should occur at the system, subsystem, or component level. The costs and benefits of commonality should be assessed in the particular use case prior to identifying what the best management construct would be for implementing such commonality (if deemed useful). We recommend that FVL take advantage of this opportunity by including such an analysis as part of the upcoming Analysis of Alternatives. Metrics for the costs and benefits of commonality and joint management can be established; these would be two different sets of metrics. Actual data from the UH-60 and V-22 programs can be used to identify and quantify both the costs and benefits of the commonality achieved to date and the management constructs used.

We also recommend that any future FVL management construct use existing organizations, authorities, and processes to the maximum extent possible. This argues for a lead service (executive agent) construct with representation from other participating services as appropriate. The attributes and enablers of successful joint management discussed earlier should also be followed as much as possible. Again, the current FVL construct of the ESG, JCoC, IPTs, and JCIC and its current operating principles (e.g., voluntary participation, active planning and coordination, open lines of communication) offer a solid foundation to build on.

Acknowledgments

We gratefully acknowledge the time and accumulated wisdom we received from the many people we spoke with in the course of this study. COL Kevin Christensen, Deputy Director J-8, represented the joint sponsorship of this study and provided insight into many of the issues we needed to address. Tom Faupel, from his role in the Joint Coordination and Integration Cell, provided an excellent point of contact for us and facilitated our interactions with the larger Future Vertical Lift (FVL) community. All of the FVL stakeholders we spoke with were generous with their time and candid with their comments. We want to especially acknowledge and thank Col Rob Freeland, COL Ramsey Bentley, and Sean Faubion.

Many of our RAND colleagues provided comments and feedback on our ideas, in addition to identifying other issues to address and asking the hard questions. We offer our thanks to Elliot Axelband, Obaid Younossi, Cynthia Cook, Laura Werber, Bruce Held, Amado Cordova, Brynn Tannehill, and Bill Shelton for being generous with their time and their feedback. We are especially appreciative of the thoughtful peer reviews provided by Amado Cordova and Mark Lorell. The resulting report is much improved.

Any remaining errors of omission or commission are the responsibility of the authors.

CHAPTER ONE

Introduction

Background

In May 2008, the Secretary of Defense directed the Office of the Under Secretary of Defense for Acquisition, Technology and Logistics (OUSD[AT&L]) and the Joint Staff to examine a joint approach to the development of Future Vertical Lift (FVL) aircraft for all the military departments. The fiscal year (FY) 2009 National Defense Authorization Act called for this effort as well.[1] The FVL initiative grew out of these directives. In May 2015, the RAND National Defense Research Institute was tasked with examining alternative joint management constructs suitable for FVL.

Joint management of major defense acquisition programs (MDAPs) has often been posited as a way to reduce redundancy and duplication across the services; avoid duplicate investment in technologies and systems; and generate savings in the production, operations, and support phases of a given program. FVL is one of the more prominent recent examples of the U.S. Department of Defense (DoD) attempting to achieve such cost savings and additional efficiencies through multiservice collaboration or joint management on complex programs.

In the DoD context, *joint program management* means that two or more services are working together to achieve a shared objective. In that sense, it is both an operational and a management construct.

---

[1] Deputy Secretary of Defense, *Future Vertical Lift: A Strategic Plan for United States Department of Defense Vertical Lift Aircraft*, Version 2, October 1, 2011.

Commonality is a process or a technical or engineering property of a physical (hardware or software) item—part, component, subsystem, or system—that is used in multiple weapon systems or by multiple services. The two concepts are related but are not the same; commonality is often the mechanism through which joint program management yields benefits. Commonality for its own sake is rarely warranted; commonality selectively applied to systems and subsystems with similar requirements is more likely to yield long-term benefits. It is widely accepted within the defense acquisition community that joint program management and commonality go hand in hand; we argue in this report that the benefits of commonality can be achieved through management constructs that are not strictly joint.

As a matter of policy and management philosophy, DoD encourages joint management to achieve commonality. As noted in a statement by Under Secretary of Defense for Acquisition, Technology and Logistics Frank Kendall:

> The effort to eliminate redundancy across portfolios is a work in progress. It demands vigilance and constant attention to the possibilities for efficiencies by all parties. Three examples from my experience of the last two years are the Air Force Space Fence and Navy AMDR [Air and Missile Defense Radar] programs, the USMD Gator radar and the Air Force 3DLR [3-D Long Range Radar] program, and the USMC [U.S. Marine Corps] and Army light tactical vehicle programs. In each case, I have initiated or supported efforts to eliminate redundancy at system or component levels. This is largely a matter of consistent and continuous management attention, particularly as new programs and projects are proposed for initiation. If confirmed, I will continue the effort to identify opportunities for commonality within and across portfolios and I will insist that the services do the same.[2]

[2] OUSD(AT&L), "Advance Questions for Frank Kendall, Nominee to Be Under Secretary of Defense for Acquisition, Technology, and Logistics," March 2012.

As mentioned previously, and as illustrated by the above quote, the dominant motivation for joint programs is cost savings achieved through commonality. In theory, that means that savings accrue to DoD or the services as a whole, as opposed to the joint program itself.[3]

The extant version of Department of Defense Instruction (DoDI) 5000.02 (2015) has two paragraphs on joint program management:

> b. Joint Program Office Organization
>
> (1) A Joint Program Office will be established when a defense acquisition program involves the satisfaction of validated capability requirements from multiple DoD Components and/or international partners, and is funded by more than one Component or partner during any phase of the acquisition process. In most joint programs, a lead Component will be designated to manage the acquisition process and act as the acquisition agent for the participating DoD Components. The participating Components, those with a requirement for the program's products, support and participate with the lead DoD Component in managing the acquisition process. Joint programs will be managed in accordance with the provisions of a memorandum of agreement, and with the lead DoD Component's acquisition procedures and acquisition chain of command, unless directed otherwise by the DAE [Defense Acquisition Executive].
>
> (2) DoD Components will neither terminate nor substantially reduce participation in joint MDAP and Major Automated Information System (MAIS) programs without capability requirements validation authority review and DAE approval. The DAE may require a DoD Component to continue some or all funding, as necessary, to sustain the joint program in an efficient manner, despite approving a request to terminate or reduce participation. Memorandums of agreement between DoD Components should address termination or reduced participation by any parties to the agreement. Substantial reduction will be determined by the

[3] Thus, a joint program that generates cost savings does not retain those savings in its budget to spend on itself. Rather, those savings are usually programmed into the budget in advance (i.e., expected savings) and are therefore used in some other part of the budget.

> MDA [milestone decision authority] in coordination with the requirements validation authority, and is defined as a funding or quantity decrease that impacts the viability of the program and/or significantly increases the costs to the other participants in the program.[4]

Beyond this, policy guidance on when joint programs are appropriate and how to best manage them is scarce. The Defense Acquisition University published a third edition of its joint program management guidebook in 2004; it contains useful information on the implications of joint programs for the requirements (Joint Capabilities Integration and Development System [JCIDS]), budgeting (Planning, Programming and Budgeting System) and acquisition (DoD 5000) processes.[5]

Theoretically, the objective of joint program management is to reduce the risks and costs of development that would be borne by an individual service by spreading those costs and risks across multiple services. In addition, increased commonality should result in economies of scale and savings that can be realized during production and support phases, and there may also be potential governance savings from joint program management and oversight mechanisms.

However, joint management introduces significant complexity in terms of developing and reconciling divergent requirements and managing and overseeing an acquisition program and its requisite processes (e.g., design, manufacturing). Additionally, joint management often increases the risk of technical complications as the interests and potentially conflicting performance requirements of the different services are traded off and balanced.

In theory, savings can also be achieved through commonality in doctrine, organization, training, materiel, leadership and education, personnel, and facilities (DOTmLPF; the "m," for materiel, is often lower case to emphasize the nonmateriel elements). For instance, if multiple services use the same aircraft, perhaps pilot training and depot

---

[4] DoDI 5000.02, 2015, p. 75.

[5] Defense Acquisition University, *Joint Program Management Handbook*, 3rd. ed., Fort Belvoir, Va.: Defense Acquisition University Press, July 2004.

maintenance could be conducted jointly, thus theoretically generating savings through reduced duplication and economies of scale. If some parts or components are shared, supply chain management for those items may also be shared. It is worth noting that although commonality is the aspect of joint program management that yields its greatest benefits, commonality itself is not entirely without risk. Moreover, some degree of commonality may be achieved by means other than a joint management construct.

Challenges of Joint Program Management

There are distinct challenges that accompany a program that is jointly managed by multiple services. To begin with, having multiple services as stakeholders requires a deliberate effort to establish a shared vision and understanding of key concepts (e.g., participation, commitment, joint management, commonality). Successful joint management necessitates incentivizing and sustaining stakeholder participation. Additionally, joint management means coordinating schedules and budgets across multiple stakeholders throughout the lengthy acquisition timeline. Similarly, it often requires navigating the oversight and approval processes of multiple services.

Program managers (PMs) face a complex incentive structure in a joint environment, and they can be burdened by constraints on their ability to manage the inevitable cost and performance trade-offs that will occur. In a joint requirements development process, stakeholders may pin their respective requirements to the highest percentiles and most extreme needs, thereby clouding the joint leaders' abilities to determine whether their true requirements are actually similar enough to justify a joint program. Nonmateriel issues can be a major impediment to both commonality and joint management. Doctrine, organization, training, sustainment practices and certifications, and concepts of operations and employment are often different enough across services that they present significant barriers. Many of these differences arise from the different missions and tactics, techniques, and procedures that have historically evolved over time within a service. Some of these nonmateriel issues are deeply embedded in service culture, making them difficult to change. Jointly managed programs have to overcome the enduring parochial tensions that exist between the services.

These aforementioned challenges of joint management are in addition to the complexities that can emerge in any acquisition program. A standard (i.e., single-service) acquisition program must align capabilities, requirements, and technical performance objectives across its stakeholders, even when all stakeholders are from the same service. There is a continuous need in acquisition programs to adapt to changes in knowledge and the strategic environment. Many acquisition programs are burdened by the problem of frequent turnover of key personnel. That is, those who write the requirements and the concepts of operations tied to a particular acquisition program often rotate out of their positions before reaching the execution or implementation stage, thereby creating a continuous cycle of renegotiation on many earlier written plans and key decisions. More generally, acquisition programs must address the friction that exists between wanting to formalize and institutionalize processes and allowing for flexibility and innovation (e.g., white space) in the acquisition process. These complexities are present in all MDAPs, although they are more pronounced and often more costly in a joint environment.

A broad review of DoD's experience with joint program management and commonality suggests that they are not the same. Joint management does not necessarily result in high degrees of commonality, and some degree of commonality may be achieved by means other than a joint management construct.

CHAPTER TWO

Methodology

Objectives

The objective of this study was to enhance management of the joint FVL initiative by recommending ways to improve internal program office structure, as well as alignment with key external organizations. As part of this effort, we were asked to review historical experience with joint program management, identify factors affecting program success, and recommend ways to apply those lessons to the FVL program management organization.

Approach

We considered a series of questions that served as the foundation for this study and analysis process. We also indicate where in the report the following questions are addressed.

1. **What is being managed?** Chapter Three describes the FVL initiative as of March 2016, and the beginning of Chapter Five defines what is being managed for purposes of this analysis.
2. **What is the intended purpose or goal of FVL?** The description of FVL in Chapter Three identifies the goals of the initiative.
3. **Who are the stakeholders, and what is their rationale for participating in FVL?** Chapter Three identifies FVL stakeholders and their roles.

4. **How do the stakeholders define key concepts, such as *joint*, *common*, and *success*?** The *FVL Strategic Plan* described in Chapter Three lays out what success may look like for the stakeholders involved and broadly defines the concepts of joint management and commonality. We define these concepts more precisely later in this chapter.
5. **What are the anticipated benefits of a joint approach to FVL?** Chapter One identified the general benefits expected from a joint management approach.
6. **Are there alternative approaches that would yield the same benefits?** These issues are addressed in Chapters Five and Six.

These questions informed our research design and information collection.

This study had three interrelated tasks. We discuss them in the following subsections.

Literature Review

Joint ventures in the public and private sectors alike have been widely studied. We looked to existing academic literature and theories centered on organizational management and strategy to inform our analysis. For the purpose of identifying organizational attributes and operating procedures that had facilitated successful joint program outcomes in the past, we reviewed several historical programs that featured joint management or multiservice participation, with the goal of understanding how joint programs were (or can be) structured internally and aligned externally, the rationale for establishing joint programs, the unique challenges the joint programs faced, and the factors that appeared to be associated with program success or failure.

Additionally, we drew on the works of organizations (e.g., the U.S. Government Accountability Office [GAO], the Defense Science Board [DSB]) that have directly and indirectly addressed the subject of joint programs. A review of joint ventures and pertinent case studies from the business sector also generated valuable lessons with applicability in the context of Joint FVL, offering useful ideas for how to distribute decision rights and structure a joint partnership. Lastly, we drew extensively on the subject-matter expertise in acquisition programs and policies, organizational theory, and workforce management of many of our RAND colleagues.

Prior RAND research has examined the structure and outcomes of joint aircraft programs, identifying joint program management as one of many factors adversely affecting program outcomes. In 2013, Lorell et al. performed a study for the Air Force examining the life cycle costs (LCCs) of joint versus single-service aircraft programs;[1] in 2011, Smallman et al. analyzed the shared modular build of warships, and the risks and benefits of joint shipbuilding; in 2001, Johnson, Hilgenberg, and Sarsfield studied the challenges of interagency space system acquisition; and in 2008, Held, Newsome, and Lewis examined commonality in military equipment and offered a framework for improving acquisition decisions.[2] Additionally, RAND researchers have performed multiple analyses of joint programs that experienced Nunn-McCurdy breaches, including the Joint Precision Approach and Landing System (JPALS) in 2015; the Joint Tactical Radio System (JTRS) in 2013; and the F-35 Joint Strike Fighter (JSF) in 2011.

The historical review of joint program internal structure and external alignment allowed us to identify joint program constructs, and to derive lessons and best practices.

Interviews with FVL Officials and Select Program Representatives

We conducted interviews with FVL stakeholders and representatives from select acquisition programs to gather practical lessons learned.

---

1 LCCs consist of research, development, test and evaluation, procurement, and operations and support.

2 Mark A. Lorell, Michael Kennedy, Robert S. Leonard, Ken Munson, Shmuel Abramzon, David L. An, and Robert A. Guffey, *Do Joint Fighter Programs Save Money?* Santa Monica, Calif.: RAND Corporation, MG-1225-AF, 2013; Laurence Smallman, Hanlin Tang, John F. Schank, and Stephanie Pezard, *Shared Modular Build of Warships: How a Shared Build Can Support Future Shipbuilding*, Santa Monica, Calif.: RAND Corporation, TR-852-NAVY, 2011; Dana L. Johnson, Gregory H. Hilgenberg, and Liam P. Sarsfield, *Policy Issues and Challenges for Interagency Space Systems Acquisition*, Santa Monica, Calif.: RAND Corporation, MR-1372-NRO, 2001; and Thomas Held, Bruce Newsome, and Matthew W. Lewis, *Commonality in Military Equipment: A Framework to Improve Acquisition Decisions*, Santa Monica, Calif.: RAND Corporation, MG-719-A, 2008.

These interviews facilitated our exposure to actual management challenges that have occurred in joint settings and strategies that have been implemented in addressing those challenges. Acquisition programs tend to be relatively unique in terms of the interplay between system and program characteristics, as well as the characteristics of the organizational environment in which they are executed. The interviews with joint program officials yielded valuable insights into these different factors and the spectrum of possible approaches to successfully managing them. In the discussions, we paid specific attention to understanding the different types of joint program constructs and how the different constructs affected a program's alignment with external stakeholders.

The stakeholder organizations we interviewed included the members of the FVL Joint Council of Colonels (JCoC), the leaders of its four Integrated Product Teams (IPTs), the FVL Joint Coordination and Integration Cell (JCIC), the Vertical Lift Consortium (VLC), the Office of Cost Assessment and Program Evaluation (CAPE), and officials in OUSD(AT&L). The participating services in the FVL initiative thus far—based on their consistent presence at the monthly JCoC meetings (attended by the RAND team during the period of the study, from June 2015 to May 2016)—include the U.S. Army (designated as the FVL Family of Systems [FoS] lead component), the U.S. Marine Corps, the U.S. Navy, the U.S. Coast Guard, and Special Operations Command (SOCOM).[3]

The interviews with officials in the FVL stakeholder community and in select joint programs allowed us to validate findings from the literature review, as well as understand the nuances associated with joint management structures and how these apply to FVL.

Evaluation of Alternative Management Structures

Based on the information assembled in tasks 1 and 2, this task examined several possible alternative program models, including different options for both internal structure and external alignment. We evaluated the pros and cons of these possible management constructs in the

---

3 The Air Force was invited to participate in FVL but did not attend any of the JCoC sessions we observed during our research (June 2015 through May 2016).

context of the unique characteristics, goals, and execution environment of the FVL program. To the extent possible, we tried to associate specific characteristics of joint program structure and alignment with specific management challenges and institutional environments. We then applied lessons from this analysis to the current FVL internal structure and external organizational alignment and recommended possible changes, taking advantage of the tailoring provisions in DoDI 5000.02 to inform our recommendations.

Analytical Framework

The historical record of DoD joint programs indicates widely varying interpretations of what a joint or multiservice initiative entails. Drawing on the historical record, we developed a series of baseline definitions, assumptions, and principles that then informed our analysis. Together, these form a framework for joint management constructs.

There is a joint acquisition cost-growth premium. Lorell et al. found that joint aircraft programs faced an acquisition cost-growth premium compared with single-service aircraft programs.[4] In other words, jointly managed aircraft programs tended to incur higher cost growth than equivalent single-service aircraft programs. The research suggests that unless the system requirements among the services are the same, joint management of complex programs is not warranted. The research also suggests a cost-benefit framework to inform joint program management decisions: The expected benefits should outweigh the expected costs of joint program management.

Joint program management and commonality are not synonymous. We distinguish joint management from commonality. Joint management is an organizational construct, whereas commonality refers to the technical, engineering, or operational qualities of components, subsystems, and systems. Joint management stems from a deliberate decision, whereas commonality can emerge organically, as there are natural incentives (e.g., financial savings, benefits of mature tech-

[4] Lorell et al., 2013.

nology) for the latter. Both joint management and commonality can be viewed along a spectrum.

The light purple model at the right of Figure 2.1 can be referred to as the leader or follower, lead service, or executive agent approach. It is mainly a single-service program with multiservice participation. The model in the middle describes a merger of preexisting programs that come together during development, usually at the direction of higher authorities (e.g., Office of the Secretary of Defense [OSD], Congress). The dark purple model at the left of the figure is the most joint on the spectrum, as it describes programs that are purely joint or "joint from inception," meaning that the joint process begins in the requirements development phase. While we have broken the framework into three parts for simplicity, it is important to understand that there are variations of each of these models; the specific rules of joint management and the items being managed can vary within each of these three general models of joint management.

Figure 2.2 shows a similar spectrum for commonality. On the right, limited commonality may take the form of similar resource use (e.g., fuel) or coordinated manufacturing (e.g., same production line

Figure 2.1
Spectrum of Joint Management

RAND *RR2010-2.1*

Figure 2.2
Spectrum of Commonality

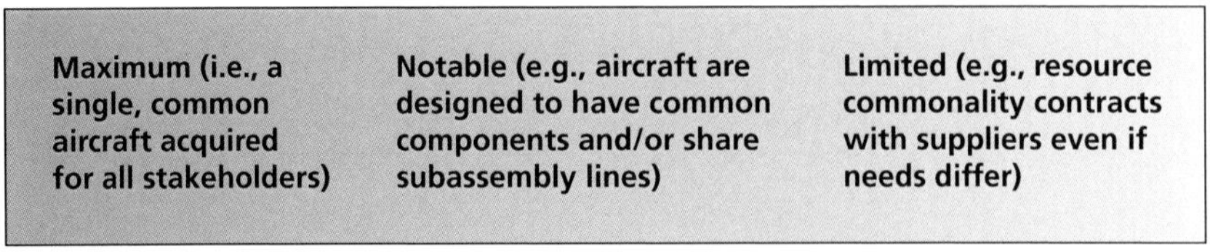

RAND *RR2010-2.2*

for basic airframe) or contracting (multiservice contract vehicle), such that multiple services are purchasing from a common supplier at some level. In the middle, services may design aircraft to use common components or to have entire subassemblies in common. This includes a case in which a service adopts a platform developed by a sister service and incorporates its own mission systems.[5] The high end of commonality (left side of Figure 2.2) would take the form of a single, entirely common aircraft procured for multiple services by a single, joint program office.[6] Commonality may also include elements of DOTmLPF—for example, common basic pilot training or a common logistics chain for a specific component.

Examples of the different forms that commonality may take include the following:

- The Defense Standardization Program is an example of an existing institutional structure and policies and procedures intended to reduce operating and support costs through "standardization of materiel, facilities, and engineering practices to improve military operational readiness, and reduce total ownership costs and acquisition cycle time."[7]
- The Navy's Seawolf and Virginia Class submarines illustrate part commonality across firms (two shipyards). The Seawolf included two design shipyards, one of which became the single construction shipyard. Many duplicate part numbers were generated during the design process, and existing design standards were not always used, resulting in a $67,834 bill of materiel parts. In contrast, the Virginia Class introduced parts standardization in the design

---

5 Another RAND research effort, conducted in roughly the same time frame but as yet unpublished, calls this *derivative commonality*. The UH-60 variants are a good example.

6 Sydney J. Freedberg, Jr., "Navy Wants to Work with Air Force on New Nukes: VADM Benedict," *Breaking Defense*, June 19, 2015.

7 Defense Standardization Program, "Policy and Guidance," web page, undated.

phase, resulting in a $27,014 bill of materiel parts, an estimated cost avoidance of $789 million over the life of the program.[8]

- Commonality may also happen within a service. The Navy's decision to use the existing LPD-17 hull form for its future LX(R) design is expected to reduce production costs through learning effects and supply chain continuity, as well as reduce the operation and maintenance costs of both ship classes.[9]

Evidence of the benefits and costs of commonality is sparse. Because of insufficient data, it remains difficult to determine whether the realized benefits of joint programs and commonality exceed or even offset the increased costs stemming from the greater complexity and nonoptimal designs of joint or common initiatives. Limited quantitative data have been collected on benefits and costs of pursuing joint management and commonality in new developments and acquisition programs.

The benefits usually attributed to joint program management (e.g., reduced LCCs, streamlined efforts in research and development [R&D] and science and technology [S&T], economies of scale in production and operations and support, consistency and continuity in business practices) may actually be attained through commonality. A joint management construct can facilitate commonality, but commonality can also be attained through different means (e.g., informal coordination or a lead service construct).

Commonality has the potential to reduce the risk, the costs, and the length of time typically required for a new product to enter the market. That said, commonality is not without risks. The failure of a common component or components across the fleet (i.e., all participating services) at the same time could greatly compromise warfighter readiness. Additionally, because commonality may limit opportunities

---

[8] Defense Standardization Program Office, *Defense Standardization Program Case Study: The Virginia Class Submarine Program*, Fort Belvoir, Va.: U.S. Department of Defense, DSP-CS-15, 2007.

[9] Terry McKnight, "Commonality Drives Savings in Shipbuilding," *U.S. Naval Institute News*, February 29, 2016.

for competition in a new program, it may contribute to a reduction in supplier diversity in the United States, effectively a contraction in the industrial base. Commonality may reduce opportunities for innovation and design flexibility across the strategic industrial base as well. Lastly, joint initiatives that require the continuous production of common components over a long period of time may find themselves dependent on obsolete parts and manufacturing processes.[10] It is important to recognize that, although these risks are within the realm of possibility, at present the data on actual costs and savings realized through commonality are extremely limited, making it impossible to assess the risks and costs of commonality with precision.

Road Map for the Remainder of the Report

Chapter Three provides a brief description of the joint FVL initiative, with an emphasis on organizational structures, early planning activities, and current status (as of March 2016). This discussion provides insight into what needs to be managed going forward.

Chapter Four draws on select elements of the broad academic literature on organizational design, management practices, and change management to identify attributes of organizations relevant to joint management. It also provides several examples of joint management in the private sector, which offer lessons for structuring and executing joint management organizations in DoD.

Chapter Five provides a brief summary of DoD's experience with joint programs. We include policy statements on joint management as well as a summary of historical experience with joint program management. We provide short summaries of a low number of joint or multiservice programs that illustrate different organizational designs and

---

[10] "Joint programs that extend continuous production of common components over a long period of time risk creating future dependencies on obsolete parts and manufacturing processes," the report states. "The key challenge is sustaining critical design, engineering, and production capabilities over time." See Justin Doubleday, "Air Force, Navy Team Examining Commonality for Future Strategic Missiles," *Inside Defense: Inside the Air Force*, January 21, 2016.

other attributes of joint management. We also summarize two older studies that identify the attributes of successful joint management endeavors—lessons that are still relevant today.

Chapter Six summarizes the attributes of joint management structures that appear to enable success. We then describe alternative management constructs on both ends of the joint management spectrum. Using those aforementioned key enablers of effective joint management as our benchmarks, we assess the pros and cons of these different constructs and define additional management constructs along the joint management spectrum.

Chapter Seven pulls together our findings and draws inferences for the FVL initiative. We make several recommendations for the FVL stakeholder community to consider as the initiative moves forward.

CHAPTER THREE

Description of the Joint FVL Initiative

Origins and Current Management Construct

Given the direction by OSD and Congress in 2008, OUSD(AT&L) Land Warfare and Munitions (LW&M) and the Joint Staff began planning the FVL initiative. In September 2012, the Deputy Secretary of Defense approved the *FVL Strategic Plan*, which formally established the FVL initiative and a management construct (an organizational design with roles and responsibilities delineated) to guide implementation.[1] The *FVL Strategic Plan* itself was based on several prior studies that together provided an analytical basis for initial FVL planning activities.[2]

From a capabilities perspective, the goal of FVL is to deliver next-generation innovations in vertical takeoff and landing (VTOL) to the joint warfighter. From a management perspective, the FVL initiative is intended to enable coordination and collaboration across DoD and other government agencies (i.e., the Coast Guard, other Department of Homeland Security agencies). This goal includes joint development of concepts of operations and employment and requirements, the coordi-

---

1 Deputy Secretary of Defense, 2011.

2 The prior studies were the *FVL Capabilities Based Assessment (CBA)*, the *FVL Science and Technology Plan*, and the *DoD Study on Rotorcraft Survivability* (see Deputy Secretary of Defense, 2011, p. 9; OUSD[AT&L], *FVL Capabilities Based Assessment [CBA]*, presentation at AHS International Specialists Meeting, Unmanned Robot Systems, January 21, 2009; AHS International, "Future Vertical Lift," web page, undated; and U.S. Department of Defense, *Study on Rotorcraft Survivability: Report to Congress*, Washington, D.C., October 5, 2009).

nation of science and technology activities, the sharing of the resulting information and technologies, and acquisition planning.

A 2014 update to the *FVL Strategic Plan* states:

> The Department of Defense will design, develop and field a fleet of third generation air vehicles that will ensure the United States' dominance in the vertical lift domain throughout the 21st century and beyond. The Department will aggressively pursue the most capable aircraft at the best value by minimizing development, acquisition, and [LCCs] through Joint solutions of common core technologies, architectures, and training, emphasizing the ability to conduct safe, reliable and continuous operations world-wide in all environmental conditions.[3]

In its study request, the JCoC referenced key priorities and expectations in the realm of commonality: "Provide for the most commonality of aircraft and system design, manufacture and support where it makes sound sense (and not do commonality for its own sake)."[4]

The FVL initiative's management construct as of March 2016 is illustrated in Figure 3.1. While the Army is designated as the lead agency, OUSD(AT&L) is the MDA to help maintain a joint program management orientation. An executive steering group (ESG) was established as the policymaking and decisionmaking body for FVL. The ESG is co-chaired by the Deputy Director, LW&M, OUSD(AT&L) and the Deputy Director, Force Management, Application and Support, J-8, Joint Staff. Participation is at the 2-star/SES level and includes all relevant stakeholders. The ESG is tasked with executing the *FVL Strategic Plan*; the ESG does not manage requirements or acquisition processes, and it is not a U.S. Code (U.S.C.) Title 10 body.[5] The ESG makes recommendations to the services. Because the members of the ESG repre-

---

[3] Deputy Secretary of Defense, *Future Vertical Lift: Report on the Strategic Plan for the United States Department of Defense Vertical Lift Aircraft*, Washington, D.C.: U.S. Department of Defense, 2014, p. 1.

[4] Deputy Secretary of Defense, 2011, p. 2.

[5] 10 U.S.C. is the law giving the military departments authority, and OUSD(AT&L) also has 10 U.S.C. authority through delegation by the Secretary of Defense.

Figure 3.1
FVL Initiative Current Management Construct

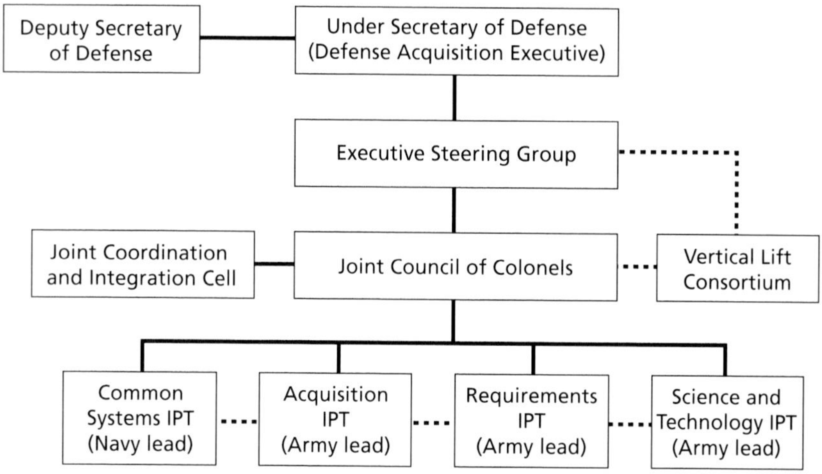

SOURCE: Adapted from AHS International, FVL overview briefing at Forum 71, Virginia Beach, Va., May 2015.
RAND *RR2010-3.1*

sent the services' 10 U.S.C. requirements and acquisition communities, the appropriate service office or agency is normally represented within the ESG structure. Thus, the ESG facilitates improved communication and information-sharing across the services, and it acts as a forum for both formal and informal joint decisionmaking regarding FVL.

The FVL JCoC is the action arm of the ESG and is charged with execution and implementation of the *FVL Strategic Plan*. The JCoC consists of colonel-level/O-6 representatives for each of the services, SOCOM, the Joint Staff, and the OSD staff. These representatives speak for their services or agencies and represent their positions on FVL matters. In its role as the action arm of the ESG, the JCoC can provide guidance and direction to the JCIC, FVL working groups, and the IPTs.

The JCIC did not develop as initially laid out in the *Strategic Plan*, because of resource limitations and the potential duplication of capabilities resident in services' requirements and acquisition organizations.[6] The JCIC is responsible for coordination, communication, and synchronization throughout the FVL community.

The ESG has chartered four IPTs:

- The Common Systems Integrated Product Team (CS IPT) enables integrated warfighting capabilities through application of common standards and processes, integration of common systems and components when appropriate, and common interface provisions for rapid implementation of unique capabilities when required.
- The Requirements Integrated Product Team is responsible for developing joint requirements based on information from all participating services. This IPT is also responsible for creating the capability-related documents needed for the requirements and acquisition processes.
- The Science and Technology Integrated Product Team (S&T IPT) completed the Joint Multi-Role Technology Demonstrator (JMR TD) design and initiated component tests and demonstrations (see discussion below), established working groups to address technological challenges, synchronized system-level technology development road maps with the broader FVL investment strategy, and established a relationship with industry. The S&T IPT has also been responsible for delivering system design, analysis, test, and fabrication data to support the Business Case Analysis (BCA) and Analysis of Alternatives (AoA).
- The Acquisition Integrated Product Team has focused on developing the BCA and the project management office personnel and facilities plans; has contributed to the AoA study guidance and study plan, as well as the analysis of FVL's acquisition strategy; and is responsible for preparing for and ultimately executing the Material Development Decision (MDD) Defense Acquisition Board (DAB).

---

[6] Deputy Secretary of Defense, *Future Vertical Lift: Report on the Strategic Plan for the United States Department of Defense Vertical Lift Aircraft*, Washington, D.C.: U.S. Department of Defense, 2012.

Membership for all IPTs includes the Army, Navy, Marine Corps, Air Force, SOCOM, and Coast Guard.

The *Strategic Plan* and an update to that plan[7] were careful to state that the FVL management construct does not supplant or contravene any staff, command, or service responsibility or authority as described in 10 U.S.C., public law, federal or service regulation, or DoD-established policies. The services, through their representation on the IPTs, retain their 10 U.S.C. responsibilities, and the JCIC focused on its role of coordinating and synchronizing FVL activities. This management construct was also more efficient than an expanded JCIC organization; it generated minimal resource and personnel requirements. The four IPTs took on roles originally envisioned for the larger JCIC.

Participation in the JCoC and the IPTs is an "other duty as assigned," meaning that none of their constituents is focused on the JCoC or FVL as a principal role. However, FVL-related activities do fall within the scope of the primary official role within their respective services or organizations. The VLC is a private-sector organization composed of both industry and academic members whose mission is "to work collaboratively with the U.S. Government to develop and transition innovative vertical lift technologies to rapidly and affordably meet warfighter needs."[8] The JCoC has established a formal relationship with the VLC through an alternative transaction authority mechanism. The VLC has provided analysis and recommendations to the JCoC through a series of papers and attendance at select meetings.[9]

---

7 Deputy Secretary of Defense, 2014.

8 VLC, presentation briefing to the Army Science Board, July 24, 2015b.

9 See for example, VLC, "The Vertical Lift Consortium Views and Recommendations for the Future Vertical Lift Strategic Plan," unpublished white paper, 2015; VLC, briefing presentation to the Army Science Board, July 24, 2015.

The Strategic Plan and Current Status

The FVL initiative management construct described above is responsible for the management and execution of the *FVL Strategic Plan* and for synchronizing the efforts of the services and agencies involved in the effort. Key functions under the *FVL Strategic Plan*, such as requirements development, program management, programming, budgeting, manning, and equipping, still reside at the service, SOCOM, or appropriate staff level.

The *FVL Strategic Plan* focuses the FVL stakeholder community on joint requirements development and early planning. The FVL community produced a joint initial capability document (ICD) in April 2013.[10] The ICD broadly outlines a full range of potential mission needs and desired capabilities for future rotorcraft. An MDD was planned for October 2016, followed by a two-year period to conduct a joint AoA. Most of the activity to date has focused on joint development and the grouping of concept of operations (CONOPS) and concept of employment, identification of affected mission areas across the stakeholder community, and the development of joint requirements and capability sets that are roughly consistent with size and weight parameters (light, medium, heavy). This information is then used to derive sets of performance specifications for each capability set. Figure 3.2 summarizes the results of this set of activities through January 2016.

The JCoC has met monthly since approximately January 2015 to discuss the results of the IPTs' activities. Much attention has been given to refining the capability sets and associated performance specifications, developing a draft joint capability development document (CDD) to support MDD, and developing input to the AoA guidance that CAPE will then issue.

There is a generalized and notional schedule mostly for planning purposes; FVL was not yet (as of May 2016) a Program of Record (PoR), meaning that it has not yet formally entered the acquisition process. There is a fairly long planning and analysis phase (four years)

---

[10] U.S. Army, *Initial Capability Document (ICD) for Future Vertical Lift (FVL) Family of Systems (FoS)*, April 8, 2013, not available to the general public.

Figure 3.2
Overview of FVL Capability Sets

FVL FoS				
Light	Medium			Heavy
All air vehicles have common: Cockpit—FACE/JCA—Training—Requirements—Reduced overhead—Mission flexibility Sustaining—Maintaining—Repair parts and components—Aerial refuel				
Missions • Reconnaissance • Security/urban assault • CAS/attack • DA • Maritime interdiction ops • MEDEVAC	**Missions** • Reconnaissance • Security/urban assault • CAS/attack • SuW/ASW/MCM • DA • Maritime interdiction ops • MEDEVAC • NEO • CSAR • Rotary wing intercept • Search and rescue • Counterdrug	**Missions** • Air assault/urban assault • Amphibious assault • CAS/attack • HA/DR • Tactical resupply • SuW/ASW/MCM • DA • Maritime interdiction ops • MEDEVAC • NEO • CSAR • Aerial refuel donor	**Missions** • Air assault/urban assault • Amphibious assault • HA/DR • Tactical resupply • DA • Maritime interdiction ops • MEDEVAC • NEO • CSAR • Aerial refuel donor	**Missions** • Air assault • Amphibious assault • HA/DR • Tactical resupply • DA • Maritime interdiction ops • MEDEVAC • NEO • CSAR • MCM • Aerial refuel donor
Capability Set 1 • Speed 200–250 knots • Radius 170–229 nm • Passengers 6 • Crew 2 • Int payload 2–2.5k • Ext payload TBD • DDG compatible • Self-deployment • Endurance 30 min–2 hrs @ 108 nm	**Capability Set 2** • Speed 170–270 knots • Radius 300–437 nm • Passengers 8–10 (@ 285 lbs.) • Crew 4 • Int payload 3.5–5.5k • Ext payload 6–8k • DDG compatible, SL103° • Self-deployment • Endurance 4.5 hrs @ 50 nm	**Capability Set 3** • Speed 270–350 knots • Radius 300–450 nm • Passengers 10–12 • Crew 4 (@ 250 lbs.) • Int payload 3.5–5.5k • Ext payload 6–8k • LPD/LHD compatible • Self-deployment • Endurance 30 min @ 300 nm	**Capability Set 4** • Speed 250–300 knots • Radius 324–420 nm • Passengers 24–32 • Crew 4–2 • Int payload 12–20k • Ext payload 15–20k • HOGE SL/103 • LPD/LHD compatible • Self-deployment • Endurance 20 min @ 24 nm	**Capability Set 5** • Speed 270–350 knots • Radius 750–1,200 nm • Passengers 45–54 • Crew 4 • Int payload 24–30k • Ext payload 30+k • HOGE 6K/95 throughout • LPD/LHD compatible • Self-deployment • Endurance 30 min @ 750 nm

SOURCE: Adapted from JCoC, Briefing V.3, Slide 33, October 15, 2015.
NOTES: ASW = antisubmarine warfare; CAS = close air support; CSAR = combat search and rescue; DA = direct action; DDG = destroyer; DR = disaster relief; ext = exterior; FACE = Future Airborne Capability Environment; HA = humanitarian assistance; HOGE = Hover Out of Ground Effect; int = interior; JCA = Joint Common Architecture; LHD = amphibious assault ship; LPD = amphibious transport dock; MCM = mine countermeasures; MEDEVAC = medical evacuation; NEO = noncombatant evacuation operation; SL = sea level; SuW = surface warfare; TBD = to be determined. A pressure altitude of 6,000 feet and an ambient temperature of 95°F (6K/95) HOGE at the objective (midpoint) is required, and 6K/95 HOGE throughout is desired, unless otherwise stated.

RAND *RR2010-3.2*

leading to a Milestone A decision scheduled for early in FY 2021. This period includes the MDD milestone and the AoA, and it is in addition to the planning and analysis that has gone on since approximately 2009. A three-year technology maturation and risk reduction (TM&RR) phase is planned, leading to a Milestone B and entry into engineering and manufacturing development (EMD) in early FY 2024. Milestone C and low-rate initial production (LRIP) are planned for FY 2029, with initial operational capability of the first capability set notionally planned for the early to mid-2030s.

This notional schedule is also oriented toward the first of the expected series of FVL programs. The original *Strategic Plan* (2012) laid out a basic approach to requirements development and acquisition that is reflected in current plans. Defined as an FoS, FVL was never intended to become a single large acquisition program that addressed every mission and capability set for every service at the same time. Rather, the FoS concept allows for multiple programs over time, with each program associated with a specific capability set that includes only the stakeholders whose mission and capability needs are met by that capability set. This was not intended to preclude the use of common subsystems or components (i.e., engine, airframe, elements of the mission system) across capability sets or programs. The *Strategic Plan* did not specify a programmatic solution and, in fact, left ample room for a range of possible program constructs. Thus, each program could include both joint management and commonality, and there could also be commonality across programs over time. The degree of joint management and commonality was left undefined; the intent was to use the planning and coordination mechanisms of the FVL management construct to define and refine joint requirements and associated capability sets. A PoR would spin out of the FVL community when approved mission needs, validated requirements, and available funding come together. Those intentions remained unchanged as of March 2016.

The Army has funded a technology development program, known as JMR TD. The purpose of JMR TD is to "demonstrate transformational vertical lift capabilities to prepare DoD for decisions regarding the replacement of the current vertical lift fleet."[11] As of February 2016, Bell Helicopters and Lockheed Martin's tilt-rotor V-280 Valor (see Figure 3.3) and Sikorsky[12] and Boeing's coaxial SB-1 Defiant (see Figure 3.4) will both undergo flight demonstrations under the JMR contract. Although not intended as the prototype for FVL, the JMR TD program has been iterating on key design features that could inform research, development, test, and evaluation (RDT&E) activities for FVL. Phase 1 of JMR TD focuses on the airframe. Phase 2 will

Figure 3.3
Artist's Rendering of Bell-Lockheed Martin V-280 Valor

SOURCE: Bell Helicopter illustration via U.S. Army.
RAND *RR2010-3.3*

---

[11] JCoC, briefing, July 23, 2015a, p. 29.

[12] Lockheed Martin acquired Sikorsky in 2015; however, both companies said that they would continue working on the Joint Multi-Role (JMR) demonstrator with their respective teams, "with internal firewalls in place to prevent tainting the competition" (Sam LaGrone, "Navy Pondering Helicopter Future After MH-60 Seahawk," *U.S. Naval Institute News*, February 24, 2016).

Figure 3.4
Artist's Rendering of Sikorsky-Boeing SB-1 Defiant

SOURCE: Sikorsky-Boeing illustration via U.S. Army.
RAND *RR2010-3.4*

focus on a Joint Common Architecture (JCA) that could help determine which air vehicle and mission systems could be common across participating services in an FVL PoR, as well as across capability sets and PoRs over time. The JMR schedule, as of March 2016, is shown in Figure 3.5.

There are certain key characteristics of the current (as of March 2016) management construct of the FVL initiative that are worth noting:

- Voluntary participation by interested stakeholders. All the service, OSD, and Coast Guard participants fund their own participation at their desired level, with FVL as an ancillary responsibility for the individuals involved. The only full-time personnel are located in the JCIC, which includes a single support contractor and an Army LTC; the Army funds JCIC.

Figure 3.5
JMR TD Schedule

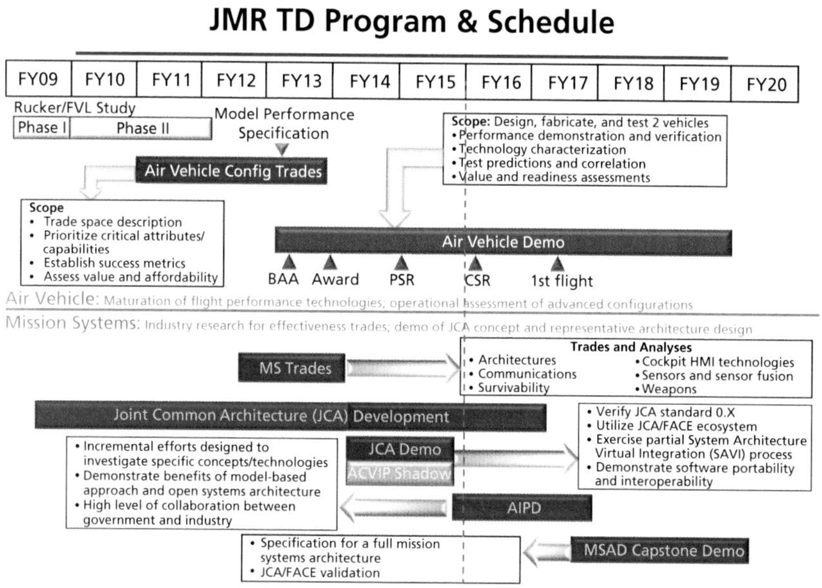

SOURCE: U.S. Army Research, Development and Engineering Command and U.S. Army Aviation and Missile Research Development and Engineering Center, "Joint Multi Role Technology Demonstrator (JMR TD) Update," October 29, 2014.
NOTE: Schedule as of March 2016. Red dashed line indicates March 2016 on the timeline. ACVIP = Architecture Centric Virtual Integration Process; AIPD = Architecture Implementation Process Demonstration; BAA = broad area announcement; CSR = critical system review; FACE = future airborne capability environment; HMI = human machine interface; MS = mission system; MSAD = Mission Systems Architecture Demonstration; PSR = preliminary system review; SAVI = System Architecture Virtual Integration.
RAND *RR2010-3.5*

- There is no real money at stake yet. There is no FVL community pool of funding supporting joint activities. FVL stakeholders largely pay for their own level of participation (except JCIC).[13]
- The current management structure has provided a reasonable set of mechanisms to share information and coordinate activities at several organizational levels (working, middle management, and senior management).

---

[13] Note that this study was jointly funded.

Future Plans

Interestingly, none of the services' aviation modernization plans mentions FVL, nor do they incorporate a placeholder for an FVL program in the out-years.[14] Rather than indicating a lack of interest or support, the absence of FVL in the services' aviation plans may be because there is no formal PoR and, therefore, no official budget line. We would expect the aviation plans to incorporate FVL as the concept matures and formal acquisition programs are established.

---

[14] Marine Aviation, *Marine Aviation Plan 2016*, Washington, D.C.: Headquarters, Marine Corps, Department of Aviation, March 2016; U.S. Army, *National Commission on the Future of the Army: Aviation Restructure Initiative (ARI)*, May 18, 2015, not available to the general public; U.S. Department of the Navy, *Naval Aviation Vision 2016–2025*, January 2016.

CHAPTER FOUR

Lessons from Organizational Theory and Academic Literature

Success will have different meanings depending on the context of a joint initiative (i.e., public versus private sector, domestic versus international). However, academic theories and business management case studies offer rich lessons that FVL may wish to draw on. There are many publications and known figures in academia and in business offering what they view as best practices in management, organizational design, and joint ventures. In the interest of brevity, in this chapter, we summarize just some observations and recommendations from these offerings and discuss their relevance to FVL. Table 4.1 highlights some of the publications we reviewed and identifies key issues and lessons that FVL should consider, irrespective of the management construct it chooses to implement.

Managing Joint Ventures

Strategic Alignment and a System of Governance

In 1991, Bamford, Ernst, and Fubini assessed the performance of 49 joint business ventures and alliances and found that only 51 percent had been "successful," with *success* defined as having each partner achieve returns greater than their cost of capital. In 2001, a decade later, their research team assessed 2,000 of what they describe as alli-

Table 4.1
Summary of Key References

Study	Key Findings
Bamford, Ernst, and Fubini (2004)	• Invest in early planning • Ensure strategic alignment across all stakeholders • Have a well-defined system of governance and terms of engagement (i.e., rules of the road), including reporting relationships • Do not impose excessive oversight and penalties that would stifle innovation • Establish rigorous metrics for performance tracking and risk management and mechanisms for accountability
Killing (1982 and 2014)	• Joint ventures can take on different constructs (e.g., dominant parent, shared management), and they all come with trade-offs • In the case of shared management, decisionmakers have to be more deliberate about preparing for and reducing risks by – developing a detailed management process before the venture is active – remaining flexible and modifying the organizational construct of the venture to meet changing needs and conditions – affording the general manager of the joint venture sufficient autonomy to make decisions and secure early successes
Gupta and Wang (2013)	• Have a focused, narrowly defined charter for the joint venture • Choose a partner that carries a low risk of conflict in the long term • Allocate decision rights (i.e., authority) based on context and logic • Build understanding and trust among stakeholders • Agree up front on a set of terms and procedures to be used in case partners decide to end the joint venture
Christensen, Hall, Dillon, and Duncan (2016)	• Focus on outcomes, not tasks • Determine what the customer needs and what the customer is seeking to accomplish • Rethink the product or service being developed and determine whether it is the right product or service to fulfill customer needs
Kotter (1996 and 2014)	• Eight-step process for leading change: – create a sense of urgency – build a guiding coalition – form a strategic vision – enlist a volunteer army – enable action by removing barriers – generate short-term wins – sustain acceleration – institute change

ance announcements and found a similar success rate (53 percent).[1] Findings published in 2004 from Bamford, Ernst, and Fubini underscore the critical importance of having strategic alignment across the separate entities involved with a joint venture and having a governance system that promotes shared decisionmaking and oversight.

> Even though a joint venture [is not] necessarily a marriage for life, governance problems can quickly trigger termination of the deal. Weak controls can cost the parent companies money and can expose them to unexpected risks. The secret to effective governance is balance: providing enough oversight to protect important assets without stifling entrepreneurship.[2]

Bamford, Ernst, and Fubini underscore the importance of having strategic alignment across all stakeholders and having a well-defined system of governance. Both of these principles can be applied in the context of FVL. Strategic alignment translates to similar requirements in the FVL context. A well-defined system of governance translates to establishing clear roles, responsibilities, and authorities for participating FVL stakeholders. Moreover, from Bamford, Ernst, and Fubini, we learn that it may serve the FVL initiative well to establish clear parameters and terms of engagement (i.e., rules of the road) for its participating services but then afford leaders the opportunity to innovate and iterate on ideas and to do so without the burden of excessive oversight and the risk of penalties for new innovative ideas that are not successful.[3]

---

1 James Bamford, David Ernst, and David G. Fubini, "Launching a World-Class Joint Venture," *Harvard Business Review*, February 2004.

2 Bamford, Ernst, and Fubini, 2004.

3 The notion of innovation as a risky, iterative process appears often in the literature on innovation and management in the public and private sectors. Robert Behn, for example, writes of the processes of "discovery, failure, learning, adaptation, some success, followed by yet more experimentation, discovery, failure, learning, and subsequent adaptation" required for innovation (Robert Behn, "The Adoption of Innovation: The Challenge of Learning to Adapt Tacit Knowledge," in Sanford Borins, ed., *Innovations in Government: Research, Recognition, and Replication*, Cambridge, Mass.: Brookings Institution Press and John F. Kennedy

Different Constructs for Joint Ventures

In a 1982 report published by the *Harvard Business Review*, Killing examines 37 joint ventures in North America and Europe and describes two possible organizational constructs for joint ventures: *dominant parent* and *shared management* ventures.[4]

The dominant parent construct, comparable to the lead service approach observed in DoD acquisition, involves a dominant parent at the top who selects all the functional managers for the venture, as well as a board of directors. Similar to the FVL JCoC, the board of directors in the dominant parent model is composed of executives from each "parent." However, the dominant parent controls all operating and strategic decision processes, leaving the board of directors with a more ceremonial role.[5] Yan and Luo mention that U.S.-Japanese partnerships in recent decades have often conformed to the dominant parent construct, with the Japanese partner typically taking on the dominant role. The joint venture between General Motors and Toyota (NUMMI) and the joint venture between Chrysler Corporation and Mitsubishi Corporation (Diamond-Star Motors) are identified as notable examples of joint ventures with a dominant parent.[6]

In the shared management model, comparable to the joint-service approach in DoD acquisition, both parents manage the venture and both contribute functional personnel. Like the dominant parent model,

---

School of Government, Harvard University, 2008, p. 158). In DoD, a risk-averse culture may impede the necessary process of innovation.

4 See J. Peter Killing, "How to Make a Global Joint Venture Work," *Harvard Business Review*, May 1982. Although he focuses entirely on *dominant parent* and *shared management* approaches in his work, Killing alludes to a third approach, *independent ventures*, which are free from interference from either parent, and notes that these types of ventures generally perform well. Researchers Yan and Luo expanded Killing's list to include five types of joint ventures: dominant control, shared control, split control, rotating management, and independent joint ventures (Aimin Yan and Yadong Luo, *International Joint Ventures: Theory and Practice*, Armonk, N.Y.: M.E. Sharpe, 2001).

5 Killing first published his book *Strategies for Joint Venture Success* in 1983; a more recent edition came out in 2014. The book contains many of the lessons offered in the *Harvard Business Review* piece (J. Peter Killing, *Strategies for Joint Venture Success*, New York: Routledge, 2014).

6 Yan and Luo, 2001, p. 92.

the shared management approach has a board of directors composed of executives from each parent, but the board of directors in this model actually functions as a decisionmaking body, similar to the FVL ESG. Yan and Luo note that partners often provide different but complementary types of resources and expertise in these types of ventures.[7] Royal Dutch Shell, Unilever, and British American Tobacco are listed as examples of successful joint ventures that have used the shared management model.[8]

The joint ventures mentioned here were formed because of a perception that they would be mutually beneficial. For example, in the case of NUMMI, General Motors and Toyota needed each other and therefore had a vested interest in making the joint venture successful; Toyota was facing import restrictions from the U.S. Congress and wanted to start making cars in the United States, and General Motors needed to innovate its products and production practices and contend with high labor costs. In the case of Royal Dutch Shell, the terms of the merger in 1907 gave 60 percent of ownership to the Royal Dutch Petroleum Company (the Netherlands) and 40 percent of ownership to the Shell Transport and Trading Company Ltd (United Kingdom). The companies wanted to merge so that they could compete with Standard Oil (United States). According to some sources, however, the nationalistic sentiments of the time would not permit a full-scale merger or takeover of either of the two companies. So they agreed on a merger and a shared management structure. Today, they are a single legal entity, headquartered in the Netherlands and incorporated in the United Kingdom. Success in both cases derives primarily from the shared perception of mutual benefits. The two different constructs relate, in part, to the nature of that benefit: GM wanted to learn from Toyota, so Toyota took a dominant role; and the two oil companies created a shared construct reflecting their largely similar desired benefit of competing with a much larger company. The lesson for FVL is that nature of the perceived benefit from commonality or joint manage-

[7] Yan and Luo, 2001, pp. 93–94.

[8] Nancy A. Hubbard, *Conquering Global Markets: Secrets from the World's Most Successful Multinationals*, Hampshire, UK: Palgrave Macmillan, 2013.

ment constructs should be identified and agreed on early in the planning stage, because the appropriate construct to use will be, in part, a function of the nature of the desired benefit.

Reducing the Risks of Shared Management

Killing offers a question that can be applied in the context of FVL: "The trade-off between using shared management and dominant joint ventures is clear-cut: Will the extra benefit of having a partner who is helping to run the joint venture outweigh the resulting disadvantages?"[9] Will the purported benefits of joint management outweigh its demonstrated costs? Killing explicitly states that the dominant parent approach is more likely to be successful but recognizes that some joint ventures—particularly those requiring continuous input from both parents (as opposed to a one-time transfer of expertise from one parent to the venture)[10]—may require a shared management approach.[11] Killing advises decisionmakers to actively prepare for and reduce the risks of shared management. His recommendations may be relevant to decisionmakers in FVL. They include (1) developing a detailed management process for the joint venture *before it is set up*, (2) remaining flexible and willing to modify the organizational construct of the joint venture to meet changing needs and conditions, and (3) affording the general manager of the joint venture sufficient autonomy to make decisions and secure early successes.

Lessons from Multinational Business Ventures

In a short piece published in 2013, Gupta and Wang offer five strategies for how to defy the low odds of success and develop a successful

---

[9] Killing, 1982.

[10] Killing (1982) argues, "If both parents' skills are necessary to the success of a joint venture, but those of one parent can readily be transferred on a one-time basis, the other parent should dominate the venture. . . . If the skills of both parents are crucial to the success of the venture, a shared management joint venture is appropriate."

[11] Note that, according to Killing (1982), "joint ventures that draw functional managers from both parents are more difficult to manage than those that do not. [However,] joint ventures that had managers from both parents performed neither better nor worse than those that did not."

joint venture.[12] The authors drew lessons and developed their recommendations based on joint ventures in the international business sector, including Malaysia-based Air Asia's failed joint venture with Japan's All Nippon Airways and its more successful joint venture with India's Tata Group.

First, Gupta and Wang argue that to improve their odds of success, leaders have to define a joint venture's charter narrowly, as doing so "provides focus, reduces complexity, and enables companies to collaborate with different partners to meet their goals."[13] The development of the *FVL Strategic Plan* and pending efforts led by the requirements IPT have afforded participating services the opportunity to define their respective requirements and develop a shared vision of FVL; continuing to narrow the requirements and their shared vision may serve the initiative well.

Second, Gupta and Wang mention that leaders should choose a partner that carries a low risk of conflict in the long term. In the case of Air Asia and the Tata Group, according to the authors, the alliance worked well because the partners were "high on complementarities and low on conflicts."[14] If partners in a joint venture have conflicting long-term ambitions, or if they anticipate competing with each other down the line, the joint venture may face a higher risk of failure. Services participating in FVL may compete with each other for missions or budget; such competition may hinder reaching agreement on requirements, cost-performance trade-offs, and other program-level decisions.

Third, Gupta and Wang recommend allocating decision rights (i.e., authority) based on context and logic. That is, the joint venture should clearly define who has decision power in different functional areas (e.g., R&D, operations, and human resources). In the case of All Nippon Airways and Air Asia, All Nippon Airways ceded control to Air Asia on matters like customer service. The authors argue that this may not have been a logical decision, given the significant differences

---

[12] Anil Gupta and Haiyan Wang, "Beat the Odds in Cross-Border Joint Ventures," *Harvard Business Review*, October 9, 2013.

[13] Gupta and Wang, 2013.

[14] Gupta and Wang, 2013.

in their customer service practices and the expectations of their respective client bases.[15] As the different programs within FVL take shape, negotiations over the distribution of decision rights should consider the existing strengths and assets of all participating services.

Fourth, the authors emphasize the importance of building understanding and trust among stakeholders. They write: "All joint ventures are mixed motive games;[16] value creation requires cooperation while value capture requires focusing on what's best for one's shareholders. . . . An excessive or premature focus on value capture will leave them fighting over the crumbs instead of striving to make the pie bigger."[17] Leaders in joint acquisition programs often face the same "mixed motive game"; it remains the task of the JCoC and its leadership to continue to address entrenched belief systems, promote cooperation, and reduce the competition among stakeholders in the interest of a successful, cost-effective, and sustainable FVL program.

Lastly, the authors recommend that partners in a joint venture agree up front on a set of standard operating procedures to be used in case they decide to part ways and end the joint venture. FVL may wish to define the terms of separation before it stands up its programs.

Identifying the Job to Be Done

The job-to-be-done theory developed by Harvard Business School instructor Clayton Christensen, known for his work on *disruptive innovation*, argues that instead of identifying and developing the product that they want to put into market, companies have to look at products in the same way that their customers would: as a way to get a specific

---

[15] "Given the differences between the expectations of the Japanese low-cost traveler and his counterpart in the rest of Asia, it may have been smarter for ANA to have retained the final call on those decisions" (Gupta and Wang, 2013).

[16] The term *mixed motive* refers to the idea that joint ventures may involve competing or contradictory motives. To create value in a joint venture, a stakeholder must cooperate with the other stakeholders, and yet, to capture or actually derive value from a joint venture, a stakeholder must essentially focus more internally on his or her needs and shareholders.

[17] Gupta and Wang, 2013.

job done.[18] That is, companies should redefine the question that is driving the product or service they are developing; instead of asking, "What do we want to develop and put in the market?" the question should be, "What does the customer actually need?" and, more broadly, "What is the customer seeking to accomplish?"

The job-to-be-done theory draws a distinction between a product's *function* and its *job*, and this particular aspect of the job-to-be-done theory can be directly applied to the FVL initiative. Although the function of FVL may be to jointly establish a new joint helicopter program, its job, broadly speaking, is to create a family of next-generation helicopters that multiple services will use. The former is more task-oriented, and the latter is more outcome-oriented.

The inclusion of Christensen's theory is intended to demonstrate that depending on how FVL stakeholders define their job (i.e., their desired outcome), focusing on commonality may get the job done better than a joint management construct would. This recognizes that the benefits of commonality can be obtained using management constructs other than joint.

Change Management

Many of the principles underlying change management theories applied in the business management and organizational theory areas seem applicable to the joint management problem. Change management is about how to implement something new in an existing organization in a way that facilitates success. Joint program management in DoD is a significant change from the traditional single-service program model, and the activities associated with implementing change in a complex organization offer lessons for implementing a joint program in the DoD context.

---

[18] Carmen Nobel, "Clay Christensen's Milkshake Marketing," *Harvard Business School Working Knowledge*, February 14, 2011; Clayton M. Christensen, Taddy Hall, Karen Dillon, and David S. Duncan, *Competing Against Luck: The Story of Innovation and Customer Choice*, New York: Harper Business, 2016.

John Kotter, formerly of Harvard Business School, first developed his eight-step process in 1996.[19] He expanded its scope and created a new version of the eight-step process in 2014;[20] a slightly adapted version of the 2014 list, with comments tying each step to FVL, appears in Table 4.2. Kotter and his team developed these pointers to help people in positions of authority respond to and manage their organizations through periods of change. The eight steps described by Kotter can be applied in the context of FVL; indeed, some of these steps are under way, if not already complete.

Table 4.2
Application of Eight-Step Process for Leading Change to FVL

Step Number	Action
Step 1 Create a sense of urgency	**Create a significant opportunity to energize and mobilize people to contribute to your process of change.** Participation in FVL is voluntary as opposed to being mandated, meaning that participating services are likely motivated by real interest and real needs. Although industry is seeking more immediate progress, based on proposed schedule and current budgetary commitments, there is not a perceptible sense of urgency among participating services.
Step 2 Build a guiding coalition	**Assemble some sort of entity with the authority and energy to lead and sustain a collaborative change effort.** The JCoC has been established and afforded the authority to make decisions and develop a working foundation for the FVL program.
Step 3 Form a strategic vision and initiatives	**Shape a vision to help steer the change effort, and develop strategic initiatives to achieve that vision.** The *Strategic Plan* articulates a vision of FVL shared by participating services.
Step 4 Enlist a volunteer army	**Create a force of people who are ready and willing to drive change on behalf of guiding coalition.** FVL has a coalition of engaged and empowered individuals from across the spectrum of stakeholders; it is critical for these individuals to serve as evangelists and to actively support FVL as the initiative proceeds.

[19] John P. Kotter, *Leading Change*, Boston: Harvard Business Review Press, 1996.

[20] John P. Kotter, *Accelerate: Building Strategic Agility for a Faster-Moving World*, Boston: Harvard Business Review Press, 2014.

Table 4.2—Continued

Step Number	Action
Step 5 Enable action by removing barriers	**Remove obstacles to change, and address systems or structures that pose threats to achievement of the vision.** FVL leadership may have to disrupt standard operating procedures to optimize outcomes, particularly when it comes to limiting parochialism and the silos that have formed. It is important to recognize that individuals with informal authority[a] will be as important as those with formal authority.
Step 6 Generate short-term wins	**Consistently track, evaluate, and celebrate accomplishments—small or large—and tie these wins to final outcomes.** Focus on incremental progress and track data. Setting metrics for success will be a very important task in FVL, both for assessing the progress of the program itself and for creating a record for future joint programs to draw on. Early successes may help leadership win they support they need to pursue more innovative, risky ideas.
Step 7 Sustain acceleration	**Use increasing credibility (garnered through short-term wins) to change systems, structures, and policies that do not align with the vision; hire, promote, and develop employees who can implement the vision; reinvigorate the process with new projects, themes, and volunteers.** Focus on continuous adaptation and improvement. Create a learning organization that readily responds to changes (in information, threats, and operating environment).
Step 8 Institute change	**Articulate the connections between the new behaviors and organizational success, and develop the means to ensure leadership development and succession.** In addition to creating incentives for performance, demonstrate correlation between new behaviors and positive outcomes.

NOTES: To develop and sustain support for FVL, its current leadership will have to create "evangelizers" for the cause and support those with informal as well as formal authority to advocate for the direction FVL decides to take.

[a] Ronald Heifetz and his theories on adaptive leadership distinguish informal authority from formal authority. The former is rooted in individual personalities and the ability to be influential in a situation or decision process without the reinforcements of a proper title or official power, these being tools of the latter. (See Ronald Heifetz, *Leadership Without Easy Answers*, Cambridge, Mass.: Harvard University Press, 1994.)

Key Highlights of the Literature for FVL

There is a wide spectrum of publications and theories relevant to FVL, and we have summarized just a small sample that we believe have relevance to any future FVL management construct. Bamford, Ernst, and Fubini highlight the "loose-tight" model of governance and the importance of affording oversight power to those at the top of an organizational construct while still offering autonomy and authority to those on the front lines who are executing the project. They emphasize principles that emerged in multiple publications we reviewed, such as the importance of defining the terms of a venture up front, engaging in advanced planning and resolving potentially contentious conflicts before executing a project, establishing performance metrics and mechanisms for accountability, and the like. Like Bamford, Ernst, and Fubini, Gupta and Wang underline the importance of defining the terms of a joint venture—including the breakup of a joint venture—narrowly and explicitly, from the beginning. Their work on cross-border ventures pointed to the importance of allocating decision rights (i.e., the lead role on such matters as R&D, operations, and human resources) based on logic and each partner's comparative advantage. Killing's paper speaks to the pros and cons of different types of joint ventures, specifically the dominant parent model versus the shared management model, concluding that the dominant parent model is more likely to yield success in practice. Killing notes the importance of early successes to give credibility to those leading the project, a sentiment shared by Kotter. Christensen's job-to-be-done theory directs decisionmakers to focus on outcomes as opposed to tasks, to think in the bigger picture about what the customer (the warfighter) actually needs instead of simply focusing on what they are being directed to do. Lastly, in Kotter's work, we find a step-by-step strategy for creating and managing change. FVL decisionmakers would be well served by prioritizing the specific factors and forces highlighted by Kotter that enable the successful execution of a new idea or program. These authors offer a series of issues and recommendations that FVL should consider regardless of where it decides to land on the spectrum of joint management.

CHAPTER FIVE

Select Historical Examples of Joint Program Management in DoD

The history of joint programs in DoD is one of mixed outcomes. Some joint programs are considered successful (e.g., Joint Direct Attack Munition [JDAM], Joint Standoff Weapon Program [JSOW], AIM-9X, Advanced Medium-Range Air-to-Air Missile [AMRAAM]), while for others joint management may have contributed to program cost growth and schedule delays (e.g., JSF).

Vignettes and the Spectrum of Joint Management

We examined several joint programs or organizations to better understand elements of their management construct that might inform our analysis of management constructs for FVL.

We had several basic criteria in selecting programs for closer examination. First, we wanted to illustrate the different types of programs that exist across the spectrum of joint management. That is, rather than focus on programs that were purely joint (i.e., joint since inception), we decided to examine programs that demonstrated different interpretations of joint management and emphasized diversity (in construct, in business rules, and in outcomes). Second, we focused on programs that had readily accessible information and data. Therefore, we decided to focus on programs with ample public information that we could draw on (as opposed to programs that had not been as public or whose information had not been as widely published). Lastly, we selected some programs based on input from our sponsor and FVL stakeholders. In early discussions with members of the FVL initiative, we often asked stakeholders for their models of success in the context of joint management. We drew on their responses in forming our list of programs for closer analysis.

Each joint program discussed in this chapter is a short vignette focused on those characteristics offering lessons for FVL. We summarize the lessons from these vignettes, together with two historical references and a RAND study examining joint fight aircraft programs, at the end of this chapter.

Joint Integrated Air and Missile Defense Organization

To improve equipment design, maintenance, and operational collaboration among the services, the Joint Theater Air and Missile Defense Organization (JTAMDO) was established in 1997. JTAMDO was created to help develop and codify missile defense design and operational requirements specifically for in-theater systems.[1] JTAMDO was established within the Joint Chiefs of Staff organization under J-8[2] to ensure the involvement of all services.[3]

Over time, the role of JTAMDO broadened to include more than just in-theater missile defense, as it also provided support to defense exercises and analyses performed by combatant commands (COCOMs). JTAMDO's experience in defining missile defense requirements and its involvement in J-8 made it a good candidate organization to participate in a more-integrated assessment of missile defense. These changing roles led to a renaming of the organization as the Joint Integrated Air and Missile Defense Organization (JIAMDO) in the early 2000s.[4]

A primary responsibility of JIAMDO has been to lead the ongoing Joint Integrated Air and Missile Defense: Vision 2020 initiative

---

1 In-theater missile defense capabilities include Army, Navy, and Air Force technologies—e.g., Patriot and SM-3 missiles.

2 Director of Force Structure, Resources, and Assessment.

3 Geoffrey Weiss, "Seeing 2020: America's New Vision for Integrated Air and Missile Defense," *Joint Force Quarterly 76*, National Defense University Press, December 30, 2014.

4 Weiss, 2014.

under the Joint Chiefs of Staff. The Vision 2020 initiative is focused on standardizing the United States' global missile defense plan, including the development, modernization, and operation of defense equipment across DoD. JIAMDO's mission includes identifying and coordinating joint requirements for air defense, cruise missile defense, and ballistic missile defense solutions.[5]

Based on recent[6] President's Budget justification documents, approximately half of JIAMDO's budget in a given year goes to supporting analysis for multiple demonstrations or scenarios with the specific commands that work with JIAMDO. The other half of JIAMDO's budget goes to supporting working group meetings with the services, other U.S. missile defense organizations like the MDA, and NATO.

JIAMDO is a Chairman's Controlled Activity composed of military experts drawn from across the services. That special designation within the Joint Staff gives JIAMDO some oversight responsibility of the requirements process (JCIDS) for missile defense programs but no authority to direct changes in program plans or requirements. Rather, they perform and leverage study and analysis activities to help translate operational needs into requirements and capability documents. Bilateral and multilateral agreements are used as mechanisms enabling commonality. Integration is a key challenge of commonality and is critical for interoperability and interdependence among command, control, communications, computers, intelligence, surveillance, and reconnaissance (C4ISR) and missile assets (sensors, targeting, and shooting) across the joint force. JIAMDO assesses operational concepts and architectures, helping COCOMs and services define and refine air and missile defense requirements.

Since its start, JIAMDO has involved multiple services, COCOMs, and international organizations in its requirements development, planning, and analysis efforts. Funding for JIAMDO's efforts

---

5 Office of the Joint Chiefs of Staff, *Joint Integrated Air and Missile Defense: Vision 2020*, Washington, D.C., December 5, 2013.

6 Office of the Joint Chiefs of Staff, *Department of Defense FY 2013 President's Budget Submission: The Joint Staff*, Washington, D.C., February 2012; Office of the Joint Chiefs of Staff, *Department of Defense FY 2016 President's Budget Submission: The Joint Staff*, Washington, D.C., January 20, 2015.

is provided through the Joint Staff's funding line of the President's Budget as opposed to being funded by one particular service.

Joint Light Tactical Vehicle

The Joint Requirement Oversight Council (JROC) approved the Joint Light Tactical Vehicle (JLTV) program in 2006. Legacy platforms were designed prior to the widespread use of improvised explosive devices (IEDs) and similar antivehicle weapons used in more recent combat; survivability has been a key design requirement for the JLTV since the start of the program.[7]

Milestone A and entry into the technology development phase was approved in December 2007. Milestone B approval and entry into EMD for JLTV occurred in August 2012. The Army was designated the lead service, with the Marine Corps engaged from the beginning with requirements development. Development of requirements for the design phase was a joint Army–Marine Corps effort (via an IPT with involvement from multiple stakeholder organizations). The similarity of the Army and Marine Corps' operational needs also translated to their respective requirements. However, when the two services had different requirements for a particular subcomponent or system, the IPT strived to consolidate requirements. There are clear examples of compromises made during the JLTV requirements development process in the interest of commonality. The Marine Corps accepted an Army requirement for additional vehicle armor after determining that it would not interfere with its maneuverability requirements (and would improve survivability); the Army agreed to the Marine Corps' requirement for adjustable suspension that would facilitate the vehicles' placement on amphibious transport vessels, a capability that was unnecessary for the Army but had the potential to improve maneuverability over certain terrain. The Army and Marine Corps performed analyses and limited demonstrations to help convince the other party that their primary performance requirements would not be adversely affected and cost would not be significantly affected.

---

[7] Andrew Feickert, *Joint Light Tactical Vehicle (JLTV): Background and Issues for Congress*, Washington, D.C.: Congressional Research Service, RS22942, March 9, 2015.

Once the initial requirements were developed, a request for proposal (RFP) was issued in 2008 that resulted in the placement of three design contracts, one with an Oshkosh team, one with a Lockheed Martin team, and one with an AM General team. The selected vendor would manufacture all JLTVs (17,000 total, 5,500 for the Marines and 11,500 for the Army) for the first eight years of the production phase, three in LRIP and five in full-rate production. After the initial eight-year contract, the Marine Corps plans to cease ordering more JLTVs and the Army plans to recompete the manufacturing contract and rely on purchasing detailed drawings from the first contract winner to potentially allow another team to manufacture vehicles.[8]

In conjunction with Milestone C, Oshkosh was selected as the single supplier for the JLTV in August 2015. Shortly after, Lockheed Martin filed a protest with the GAO, which was resolved in December 2015 in favor of the Oshkosh down-select.

One mechanism that has been used for the JLTV program is for each joint program office (JPO) leadership position held by one service (largely the Army, as it is the lead service for the program) to have a deputy position held by another service; these deputy positions included deputy PM, deputy director of business management, deputy director of operational testing, and deputy director of engineering. Additionally, the Army and the Marine Corps each have their own program executive office (PEO) that is involved with the JLTV program for their service (PEO for Combat Support and Combat Service Support for the Army and PEO Land Systems for the Marine Corps), as well as a PM for light tactical vehicles who works with the JPO. The Army and Marine Corps agreed on the decision and approval process prior to formal program initiation. This includes using the Army's acquisition processes and function staff for most oversight.

The JLTV program was a joint program from its inception and was designed to ensure joint involvement in multiple ways. As stated above, although the Army was designated as the lead service, the Marine Corps maintained counterpart staff to all senior Army personnel in the JPO from the outset of the program, and both services maintained

---

[8] Feickert, 2015.

involvement in the requirements IPT. In addition to organizational structure and requirements involvement, the Marine Corps contributed 39 percent of research and development funding to the program and fully funded its own service-specific testing needs, including the purchase of vehicles to be used during testing. This contributed to a sense of partnership in the joint program that may not have been present if development funding was more uneven (i.e., a split based on the number of units each service planned to procure [approximately 49,000 to be procured over the life of the program for the Army and 5,500 for the Marine Corps]).[9] Contracts for the JLTV also have been managed jointly, with a single contract being placed with each supplier prior to down-select. Lastly, other services (the Navy and Air Force) have expressed interest in procuring JLTVs at a later date but have not contributed to funding the program so far and do not have their own service-specific design constraints.[10]

JLTV program cost documented in Selected Acquisition Reports (SARs) has remained relatively stable, although planned final procurement quantities have changed slightly over the life of the program. Program acquisition unit cost (PAUC) has decreased 16 percent from the initial baseline. The schedule has largely remained within threshold values, with the exception of the delay because of the work stoppage associated with the contract award protest.

Joint Precision Approach and Landing System

Preliminary Joint Precision Approach and Landing System (JPALS) development began in 1992 when OSD requested that the military, under the leadership of the Air Force, look into developing a multiservice precision approach and landing system using emerging error-correcting GPS technology. This resulted in the establishment of a joint team (the Air Force, Army, Navy, Marine Corps, and, to a lesser extent, the Federal Aviation Administration [FAA]) that developed a JPALS mission needs statement, approved by the JROC in 1995. The JPALS

---

[9] DoD, *Joint Light Tactical Vehicle*, Joint Program Office, Selected Acquisition Report, December 2015.

[10] Feickert, 2015.

mission needs statement described an all-weather, day-and-night approach and landing system with significant flexibility and interoperability such that it could be quickly deployed on any suitable landing surface around the world, including surface ships with flight decks. The desired system would also reduce reliance on operator support, other than for system deployment, and would be resistant to radio and data transmission jamming for covert use in hostile environments.[11]

In 1996, OSD approved Milestone 0 and accelerated efforts to operationalize JPALS. The system was managed as a single, joint program with the Air Force designated as the lead service until the program passed Milestone B in 2006.[12]

The JPALS program office performed an AoA to select between multiple candidate technologies. It completed the analysis in 1997, and JPALS development proceeded as planned until 2005, when the AoA was updated to account for developments up to that point. The updated document discussed multiple options that had been considered to date and narrowed the options to three. Each service selected its preferred option from the three, and although the Navy, Marine Corps, and Army reached a consensus, the Air Force advocated for a different option.

Based on internal discussions, the Navy, Marine Corps, and Army selected their preferred option—a differential GPS—because of its faster implementation time. A joint FAA-military working group was established with the primary goal of collaboration on requirements development to ensure civilian organization involvement in the JPALS organization.[13] The timing of the Air Force requirement for JPALS was closely tied to civilian approach and landing system developments, which had planned a later implementation date for their systems than

---

[11] Federation of American Scientists, "Joint Precision Approach and Landing System (JPALS) Precision Approach and Landing Capability (PALC)," 1999.

[12] DoD, *Joint Precision Approach and Landing System (JPALS) Updated Analysis of Alternatives Study Report*, November 17, 2005, not available to the general public.

[13] Military aircraft need to be equipped to land at civilian airstrips. Therefore, the FAA and military tried to synchronize their technology development and fielding activities.

the military. The preferred option of the Navy, Marine Corps, and Army was selected for all services.[14]

Once the preferred option was selected, a seven-increment implementation schedule for JPALS was agreed on by all military services involved in the program. Because the Navy's need for JPALS was most immediate, the first increment of capability was for ship-based systems and then for sea-based aircraft. The JROC approved the implementation plan for the first increment in 2007 and directed the Navy to take responsibility for JPALS at that point in the program, in accordance with the established plan to have the Air Force relinquish its responsibility as the lead organization at Milestone B, scheduled for the end of 2006. The second increment incorporated all remaining fixed and mobile system needs, including land capability, and future increments added additional capability to expand support to lower-visibility scenarios, unmanned vehicles, and special operations forces. It was expected that the program lead would change for future increments, with the lead agency determined by whose equities dominated a particular capability increment. As of 2010, the first two increments had been approved by the JROC.[15]

By early 2014, the FAA had decided to continue using the legacy Instrument Landing Systems (ILS) for commercial aircraft rather than deploying technology interoperable with JPALS. This decision effectively reduced the Army and Air Force's need for JPALS, and they withdrew from the program. It was also the primary cause of a Nunn-McCurdy unit cost breach, which led to the restructuring of the program. The Navy remained the lead for the Navy and Marine Corps efforts, which were consolidated into a single development phase (Increment 1A). This restructured program also eliminated the requirements for land-based training systems and installation on legacy aircraft, meaning that the first aircraft that would use JPALS would

---

14 DoD, 2005.

15 Jennifer Kavanagh, Megan McKernan, Kathryn Connor, Abby Doll, Jeffrey A. Drezner, Kristy N. Kamarck, Katherine Pfrommer, Mark V. Arena, Irv Blickstein, William Shelton, and Jerry Sollinger, *Joint Precision Approach and Landing System Nunn-McCurdy Breach Root Cause Analysis and Portfolio Assessment Metrics for DoD Weapons Systems*, Vol. 8, Santa Monica, Calif.: RAND Corporation, MG-1171/8-OSD, 2015.

be the F-35C in 2026, rather than the F/A-18 in 2016, as previously planned.[16] The Nunn-McCurdy breach illustrates the risk to joint programs of both requirements changes and stakeholders dropping out.

The JPALS program started as a fully joint program, but the level of joint involvement changed over time. When the JPALS program first began, it was thought that a new standard approach and landing system could be developed for use by all branches of the military and the civilian sector, with potential worldwide application. JPALS stands out among historical joint programs because of its joint management structure and timeline. Each increment of the program focused on a specific set of capabilities, and the service tied to the set of capabilities in a given increment would be its lead service; additionally, each increment would have its own series of milestone decisions and acquisition documentation. This program-level management construct is similar to that described in the *FVL Strategic Plan* and in the JCoC meetings that the RAND team attended.

Joint Tactical Radio System

During the 1990s, the lack of a cohesive plan for interoperability and obsolescence management for the military's many radio system technologies was a concern for Congress and senior DoD officials.[17] Specifically, each service relied on its own set of radio systems for different operational scenarios, and the services managed their respective system upgrades separately. With the increasing emphasis on joint and international combat operations, radio interoperability and audio and visual information-sharing were especially important to military leadership. In response to this situation, DoD established the Joint Tactical Radio System (JTRS) program in 1997, designating the Army as lead service and overall acquisition manager for the program, with a stated goal of replacing approximately 200 radio variants across all services.[18] The JPO was formally established in 1998.

---

[16] Kavanagh et al., 2015.

[17] GAO, *Defense Acquisitions: Challenges Associated with Implementing the Joint Tactical Radio System*, Washington, D.C., GAO/NSAID-99-179, September 1999.

[18] GAO, 1999.

To accommodate service-specific needs, the JTRS program was initially divided into five clusters that each dealt with a type of radio. This provided the opportunity for services to be involved in the radio system designs that they would use most frequently.[19] Commonality among these different types of radio systems would be realized through software and a common set of modular hardware that could be configured to provide the necessary capabilities for each design. The Army planned to use JTRS as the communications backbone for its Future Combat Systems, a modernization program centered on a system of systems that rely on advanced wireless networking communications.[20]

In the early 2000s, issues with technology maturity, acquisition strategy, and the fact that each service held its own contract for its JTRS clusters resulted in program cost overruns and schedule delays. Multiple GAO reports from this time period suggest that the complicated multiservice management model of the JTRS program was the source of the issues that the program experienced.[21] Because JTRS relied on funding from multiple services, its budget was vulnerable to changes in services' respective budgets, leading to fluctuations in overall program funding.

Another consequence of this organizational structure was the lack of a streamlined method for services to resolve disputes on the joint interface requirements, which led to issues designing radios that could meet both service-specific and joint interoperability requirements. In addition to requirements development issues, available hardware and software technology proved to be insufficient to meet the JTRS pro-

---

[19] Ground vehicle and helicopter radio development was led by the Army; hand-held radio development was led by SOCOM; fixed site and maritime radio development was led by the Navy; high-performance fixed-wing aircraft radio development was led by the Air Force; and handheld, dismounted, and small form factor radio development was led by the Army.

[20] Andrew Feickert, *The Joint Tactical Radio System (JTRS) and the Army's Future Combat System (FCS): Issues for Congress*, Congressional Research Service, RL33161, November 17, 2005.

[21] GAO, *Restructured JTRS Program Reduces Risk, but Significant Challenges Remain*, Washington, D.C., GAO-06-955, September 2006.

gram's technical requirements, specifically size and weight. This further delayed the program schedule until technology matured.[22]

As described in a 2006 GAO report on JTRS, the numerous schedule delays had significant operational effects as ground troops increased their presence in Iraq and Afghanistan in 2003 but continued to operate with legacy 1980s radio systems that had limited interoperability. To adequately support ongoing military operations, the Army began to purchase encrypted non-JTRS radio systems. As the Army began to find non-JTRS radio solutions, DoD directed the JTRS program to restructure under a single joint PEO (JPEO) in the Navy's Space and Naval Warfare Systems Command (SPAWAR) and combine some of the radio development efforts.[23] The JPEO and its four joint programs were also designated an acquisition pilot program and encouraged to design a unique oversight process for decisionmaking and document approval. However, this new approach to joint management lasted only a few years, after which the JPEO was dissolved and four joint programs, each lead by a separate service, were moved back into the traditional acquisition process.

Since the program restructure, the Army has continued to procure ground and handheld radios in a similar way to what was done during the early 2000s: Procure the best available commercial-style systems rather than put significant effort into developing a universal radio system.[24] The Navy and Air Force have continued to develop their multiple JTRS systems and use them as they come into service. Production contracts have been awarded for Navy-led JTRS radios, and they will begin to enter service in the 2020s.[25]

The JTRS program was a joint program from the start and has maintained some level of joint involvement over its history, although the program's structure and planned radio developments changed over

---

[22] GAO, 2006.

[23] This restructuring included a significant reduction in the number of planned communications channels and the numbers of radios planned to be procured.

[24] GAO, 2006.

[25] Data Link Solutions, "Data Link Solutions Awarded $478.6 Million Contract for Production of MIDS JTRS Terminals," June 18, 2015.

time. Issues that emerged during the requirements development process, and time-sensitive operational needs that drove services to procure new radios while JTRS was still in development, significantly changed the trajectory of the program. While the JTRS program was organized as a joint program under a single JPEO for a while, its cluster and domain subdivisions meant that it operated as a family of related programs, as opposed to a single joint program. Individual clusters operated as joint programs based on services' similar needs for one type of radio system. An example of this is the combined Navy–Air Force effort on the systems from the clusters that were combined during an early program restructure. Commonality and interoperability are maintained largely through software rather than hardware.

Joint Surveillance and Target Attack Radar System

In the early 1980s, the Army and Air Force separately pursued an airborne surveillance and targeting system, known as the Ground Moving Target Indicator (GMTI) radar, which would provide real-time long-range support to allied forces by locating enemy armored vehicles and operating bases from afar.

The original Army-specific GMTI program, known as the Stand-Off Target Acquisition System (SOTAS), which intended to use the Black Hawk helicopter as its radar-carrying platform, was canceled in the early 1980s because of cost overruns. The original Air Force–specific GMTI program, known as Pave Mover, began as a joint venture with the Defense Advanced Research Projects Agency (DARPA) and had been designed for use with the F-111, a fixed-wing aircraft that could deliver weapons in combat. Pave Mover proved successful in Air Force test operations.[26]

When it became clear that neither Congress nor OSD would fund separate GMTI programs, the Army and Air Force joined their efforts in the interest of sharing financial costs and improving operational capabilities. Thus emerged the Joint Surveillance and Target

---

[26] Richard J. Dunn, Price T. Bingham, and Charles A. Fowler, "Ground Moving Target Indicator Radar and the Transformation of U.S. Warfighting," Northrop Grumman Analysis Center Papers, February 2004.

Attack Radar System (JSTARS), with the Air Force designated as the lead service.[27] The system[28] and its requirements were developed with Army and Air Force involvement, and each service operated a specific component of the system: The Air Force operated the airborne radar and sensing element, and the Army operated the mobile ground stations used for data processing. Given that Air Force and Army had different missions, the Army had concerns that its priorities would be neglected. A memorandum of agreement (MoA) signed by the service chiefs laid out plans for how the JSTARS missions would be prioritized and helped to alleviate some of the Army's concerns.[29]

In the early 1980s, the JSTARS program designed and tested a specialized stealth aircraft, known as Tacit Blue, to carry the airborne JSTARS radar. However, based largely on cost, the services' leadership decided to deploy the JSTARS radar on a modified Boeing 707 aircraft instead of the stealth aircraft and used Army land vehicles to field the land-based data processing capability. As the 707s used for the airborne portion of JSTARS have aged, the Air Force has developed a plan to upgrade these systems to ensure that they remain in service beyond 2025. Three industry teams with three different preferred aircraft platforms[30] are currently vying for Air Force support, although a formal RFP has not yet been developed or provided to the industry.[31]

The joint management of JSTARS is not related to multiple services' using the same or similar platforms; instead, it involves the collaboration of multiple services in the development and deployment of a single weapon system. JSTARS was effectively a merger of the separate Army and Air Force GMTI radar development efforts, followed by their cooperation within a joint management construct to further develop both the technology and its operational concept.

---

[27] DoD, *JSTARS Selected Acquisition Report*, December 2001.

[28] The system was composed of sensing elements and data processors.

[29] Dunn, Bingham, and Fowler, 2004, p. 12.

[30] Northrop Grumman (responsible for current JSTARS) teamed with Gulfstream, while Lockheed Martin teamed with Bombardier and Boeing (proposing use of a 737 variant).

[31] Lara Seligman, "As Teams Vie for JSTARS, Raytheon Stays Above the Fray," *Defense News*, September 12, 2015.

V-22

As early as the 1970s, various branches of the U.S. military desired an air platform that could balance the vertical takeoff and landing flexibility of a helicopter with the over-land speed of an airplane. This interest resulted in the establishment of a NASA tilt-rotor experimental aircraft program in 1971—the XV-15. Bell Helicopter and Boeing were awarded contracts to further develop tilt-rotor technology beginning in 1972.[32]

Early tilt-rotor technology development heavily relied on Army funding. In 1981, DoD initiated a joint effort with the Army, Navy, Marine Corps, and Air Force to design and manufacture a tilt-rotor aircraft, with the Army as the lead service, known as the Joint Vertical Launch Vehicle (JVX). Operationally, JVX was meant to serve as an electronic warfare platform for the Army, a search-and-rescue and special operations platform for the Navy and Air Force, and a replacement for a canceled Marine Corps heavy transport helicopter development program.[33]

A 1982 memorandum of understanding (MoU) codified the different needs of the services, and established the Marine Corps variant as the baseline and all other models as variants. In late 1982 and early 1983, after the MoU had been signed, OSD passed program management responsibilities for the JVX program to the Department of the Navy. Shortly thereafter, in mid-1983, the Army decided to cease development of its own variant and purchase the Marine variant. In a similar time frame, the Air Force decided to reduce its planned purchase of JVXs by more than half, shifting the majority of planned procurements to the Marine Corps. By early 1986, the program was named the V-22 Osprey and entered full-scale development. [34]

In 1987, the Army decided to cancel all planned V-22 purchases based on its changing funding priorities and questions from OSD

---

[32] Al Moyers, "The Long Road: AFOTEC's Two-Plus Decades of V-22 Involvement," Air Force Operational Test and Evaluation Center, August 2007.

[33] Jeremiah Gertler, *V-22 Osprey Tilt-Rotor Aircraft: Background and Issues for Congress*, Washington, D.C.: Congressional Research Service, RL31384, March 10, 2011.

[34] Moyers, 2007.

about the maturity and superiority of tilt-rotor technology relative to conventional helicopter technology. In addition, after the first flight of a V-22 prototype in 1989, multiple V-22 prototypes crashed during testing in the early 1990s, including one incident that resulted in seven fatalities. Throughout the 1990s and early 2000s, the changing political environment and additional crashes caused support for the program to shift dramatically.[35] The program was restructured twice; the final acquisition strategy (ultimately executed) focused first on development of the Marine variant, with the Air Force special operations variant developed later. More recently, the Navy has adopted a V-22 variant for its ship-to-shore connection fleet.

The early 1980s JVX program—the precursor to the V-22—began as a joint program that included the Army, Navy, Air Force, and Marine Corps, although the degree of joint involvement changed over time. OSD initiated the joint service agreement that led to the design and development of a single platform intended to meet multiple services' needs. Each service had its own distinct set of operational requirements for the platform. According to GAO, many of the early technical challenges experienced by the V-22 were due to reconciling the different requirements of participating services.[36]

The Air Force and Marine Corps jointly managed contracts for all V-22 variants. Sometimes there were separate contracts for subcomponents (e.g., engine, airframe), but contracts for both aircraft variants (e.g., CV-22 for the Air Force and MV-22 for the Marines and Navy) were kept together. This is demonstrated through program SARs, which show a single contract for all V-22 variants. While the multiple airframe variants are managed jointly, the engine is maintained on a separate contract from the airframes. The V-22 engine contract is also documented by V-22 program SARs.[37]

In addition to the aforementioned technical issues, cost growth has proved to be a notable issue for the V-22 program. A 2009 GAO

---

[35] Gertler, 2011.

[36] GAO, *The V-22 Osprey—Progress and Problems*, Washington, D.C., NSIAD-91-45, 1990.

[37] This information is derived from DoD's V-22 SARs from various years.

report states that as a result of technical issues and procurement quantity variations, unit procurement costs have increased almost 150 percent since the start of the program, and research and development costs have increased by more than 200 percent. A review of V-22 SARs indicates that costs have increased considerably, while quantity has been reduced by almost half from the development baseline. Recurring readiness-related issues were also indicated in historical SARs.

The V-22 program provides an example of at least two issues that should be of importance to FVL stakeholders. First, the degree of joint participation may change during program execution as a result of changing priorities or circumstances in individual services. Second, joint program management can increase the technical complexity of a program, as the services' different requirements are accommodated. This increase in technical complexity is in addition to the increase in management complexity entailed in multiservice endeavors.

That said, the V-22 also represents a model of joint management in which a basic system design forms the basis for variants for other services. This type of joint management requires multiservice participation up front to reach agreement on the basic variant, with the services expecting to fund their own mission-specific modifications and equipment. The use of an MoU capturing this agreement early in the program is notable.

Joint Primary Aircraft Training System

Prior to the Joint Primary Aircraft Training System (JPATS) program, the Air Force and Navy each had relied on separate platforms for basic pilot training. Versions of each service's platform, the T-37 for the Air Force and T-34C for the Navy, had been in service since approximately 1950, and each service desired an upgraded platform. Because of similarities in requirements, the services merged their respective efforts to develop new training platforms and decided to base JPATS on a commercial aircraft design. In some cases, the services compromised on specific requirements to facilitate maintaining a common design.[38]

---

[38] GAO, *Acquisition Plans for Training Aircraft Should Be Reevaluated*, Washington, D.C., GAO/NSAID-97-172, September 1997.

One example of a requirements compromise was the cockpit seating arrangement. Previous Air Force training aircraft utilized a side-by-side seating arrangement, with the trainer and trainee sitting next to each other in the plane's cockpit. However, the Navy preferred a tandem arrangement, with the trainer sitting in a seat behind the trainee within the cockpit. A tandem arrangement was agreed on by both services. The joint requirements document and mission needs statement were approved by the JROC in 1991, following approval of the JPATS master plan in 1989. At Milestone 0 in January 1993, the Air Force was the lead service for the JPATS program.[39]

In 1994, the JPATS RFP was sent to potential suppliers for consideration, but it was modified in 1995 based on changes to the Navy's desired quantity of aircraft. Similar ordering quantity modifications occurred throughout the 1990s as each service's training requirements changed. The JPATS program managed this by revising the total required number of aircraft on a single contract, with the supplier and the Navy–Air Force funding split being managed internally by the government. Details of the selected funding split between services can be seen in multiple JPATS SARs dating back to the start of the program.

In 1995, the Air Force formally announced the decision of a Navy–Air Force joint down-select board to award Raytheon with the JPATS design and production contract. Shortly after this announcement, the two suppliers that were not selected filed protests based on their perceptions that down-select criteria were not appropriately applied. These formal protests required a detailed GAO review, which concluded that the down-select had been performed fairly, allowing the award of the JPATS contract. Following the award of the JPATS contract, JPATS was formally named the T-6 Texan II.[40]

Delivery of the first T-6 occurred in 2002, and official use of the aircraft for pilot training began in 2003. Aircraft deliveries met

[39] See Mark V. Arena, Irv Blickstein, Dan Gonzales, Sarah Harting, Jennifer Lamping Lewis, Michael McGee, Megan McKernan, Charles Nemfakos, Jan Osburg, Rena Rudavsky, and Jerry Sollinger, *DoD and Commercial Advanced Waveform Developments and Programs with Multiple Nunn-McCurdy Breaches*, Vol. 5, Santa Monica, Calif.: RAND Corporation, MG-1171/5-OSD, 2014.

[40] Arena et al., 2014.

the required production schedule through the 2000s, although there were some schedule delays in 2004 and 2005 because of some technical issues with delivered aircraft; these issues also resulted in delays to training programs. In 2006, design changes to resolve the technical issues were implemented, and production and training both returned to planned levels. JPATS incurred a Nunn-McCurdy breach in 2006; however, the breach was largely due to a change in the rules governing program rebaselines and not from the joint management construct.[41] The program experienced a series of technical problems during program execution leading to cost and schedule growth; however, joint program management does not appear to have been a significant factor.[42] Since the 2006 restart, the T-6 has maintained operational availability for Air Force and Navy pilot training without significant production delays.[43]

JPATS began when the Navy and Air Force signed an MoA that codified cooperation in development of requirements for the system. The services actively learned about each other's training needs by visiting training bases and attending conferences. Existing documents on requirements being worked within each service were merged to form the JPATS requirements documents. There was an O-6 level working group (similar to the FVL JCoC) to facilitate cooperation. The program also initiated a joint priority list (JPL), which established rules that both the Navy and the Air Force could use to provide direction or guidance.[44]

JPATS began and has operated as a fully joint program since its inception. The platform commonality agreed on at the start of the pro-

---

41 Arena et al., 2014.

42 See Arena et al., 2014. Other research conducted within RAND's Project AIR FORCE also identified significant cost growth caused by the scale of the redevelopment and redesign effort, requiring the renegotiation of a fixed-price production contract (communication from Mark Lorell, October 2016).

43 The information was derived from DoD's JPATS SARs from various years.

44 Bill Kinzig and Dave Bailey, *T-6A Texan II Systems Engineering Case Study*, Wright-Patterson Air Force Base, Ohio: Center for Systems Engineering at the Air Force Institute of Technology, 2010.

gram and the willingness to compromise on certain technical decisions enabled and sustained the successful joint management effort. Additionally, agreement on system requirements allowed both services to purchase the same final product on one contract, possibly leading to cost efficiencies as well. Pilot training is driven by service culture and mission requirements. This program illustrates that different services can potentially collaborate on nonmateriel aspects of an acquisition.

Joint Direct Attack Munition

Based on lessons learned from Operation Desert Storm in the early 1990s, Air Force and Navy leadership were interested in developing an inexpensive, precision-strike bomb capability that would be effective in adverse weather conditions. Initial efforts to develop this capability were through separate Air Force and Navy programs started in the early 1990s that were focused on upgrading existing nontargeting warheads by adding an improved targeting capability via a peripheral device. Shortly after these programs began, Congress mandated that they be combined as a cost-savings measure.[45] The Air Force was designated as the lead service for the new JDAM program.

The JDAM program emerged from a congressional mandate requiring the Air Force and Navy to work together on developing the precision-strike bomb capabilities that they had both been pursuing separately. The services were interested in developing similar capabilities; therefore, the process of negotiating and agreeing on requirements was not a major challenge during the JDAM design phase. Similarities in the services' needs and requirements allowed the use of the same JDAM kits by each service on most of their respective fighter and bomber aircraft, greatly simplifying JDAM procurement. Currently, all Air Force and Navy JDAM procurement is done through a single contract that exists between the Air Force and Boeing.[46]

---

[45] GAO, *Precision Guided Munitions in Inventory, Production and Development*, Washington, D.C., GAO/NSAID-95-95, June 1995.

[46] Their similar capabilities include the B-52H, B-2A, B-1B, F-16C/D, F/A-18A+/C/D/E/F, F-15E, A-10C, AV-8B, and F-22; see DoD, *Joint Direct Attack Munition (JDAM)*, Joint Program Office, Selected Acquisition Report, December 2014.

In the early design phase, the per-unit cost estimates for JDAM equipment exceeded the program office's cost targets by more than 70 percent.[47] Since JDAM procurement cost was a major congressional concern from the start of the program, the program office dedicated significant effort to acquisition cost control measures. Through engagement with various industry organizations, the program office identified best practices from development and acquisition processes in the commercial sector. Ultimately, their approach included using government and industry IPTs to foster understanding between the parties, a three-stage rolling down-select process with feedback to the suppliers at the end of each stage, allowances to use commercial products, performance-based requirements rather than general specifications, and streamlined oversight processes.[48] The use of commercial procurement practices may have contributed to a significant reduction in the per-unit costs projected by the two vendors participating in the JDAM program design phase.[49]

One factor contributing to the success of JDAM was having a clearly articulated and stable set of requirements. Communication between the program office, joint user community, OSD, and the Air Force as lead service facilitated an agreement on seven requirements: weather capability, accuracy, in-flight retargeting, warhead compatibility, carrier operability, aircraft compatibility, and average unit production cost of no more than $40,000.[50]

The major successes of JDAM were largely because it was a defense acquisition pilot program implementing the many acquisition reform

---

[47] $68,000 versus $40,000; see Dominique Myers, "Acquisition Reform—Inside the Silver Bullet: A Comparative Analysis—JDAM Versus F-22," *Acquisition Review Quarterly*, 2002.

[48] Boeing Company, *Implementing Acquisition Reform: A Case Study on Joint Direct Attack Munitions*, Fort Belvoir, Va.: Defense Systems College Management, July 1998.

[49] $14,000 per unit versus the projected $40,000 per unit cost; see Boeing Company, 1998.

[50] Mark A. Lorell, Leslie Adrienne Payne, and Karishma R. Mehta, *Program Characteristics That Contribute to Cost Growth: A Comparison of Air Force Major Defense Acquisition Programs*, Santa Monica, Calif.: RR-1761-AF, 2017; Mark A. Lorell, Julia F. Lowell, Michael Kennedy, and Hugh P. Levaux, *Cheaper, Faster, Better? Commercial Approaches to Weapons Acquisition*, Santa Monica, Calif.: RAND Corporation, MR-1147-AF, 2000.

initiatives in effect in the 1990s.[51] "The focus on cost, the use of broad mission requirements, the emphasis on cost/benefits tradeoffs, the lack of Mil-Spec requirements, and the control of the contractor over configuration and technical solutions for the JDAM program produced some dramatic results."[52] JDAM consistently met or exceeded cost and schedule performance beginning during the design phase with the program's major per unit cost reduction.[53] Procurement cost changes documented in program SARs since 2010 are almost entirely due to increases in the quantity of ordered JDAM kits and not from production or procurement issues.

JDAM's designation as an acquisition pilot program, which allowed the program to tailor acquisition processes and gave the PM more flexibility and decision authority, may have contributed to mitigating any adverse effects from the joint management construct. The similarity in requirements across services, as well as the relative maturity of the technologies, also facilitated its success.

UH-60

Development for a new Army helicopter began with a 1972 RFP under the name Utility Tactical Transport Aircraft System (UTTAS). The goal of the UTTAS program was to replace the aging Bell UH-1 Iroquois, known as the "Huey," with a similarly multipurpose design that could be maintained and operated at a lower cost than the Huey. In addition to improved cost-effectiveness, other main UTTAS program goals included improved platform availability, reliability, survivability, and mission performance (e.g., speed, cargo capacity).[54] The Army awarded separate contracts to General Electric for engine design and to Boeing and Sikorsky for competing airframe designs. Contracts for the airframes involved the construction of prototypes and a series of tests,

---

[51] Lorell et al., 2000, pp. 137–192.

[52] Lorell et al., 2000, p. 150.

[53] Cynthia Ingols and Lisa Brem, *Implementing Acquisition Reform: A Case Study on Joint Direct Attack Munitions*, Fort Belvoir, Va.: Defense Systems Management College, July 1998.

[54] GAO, *Status of the Utility Tactical Transport Aircraft System Program*, Washington, D.C., PSAD-77-31, February 25, 1977.

culminating in a December 1976 down-select and contract award to Sikorsky for the first UH-60A Blackhawk helicopters.[55]

In parallel, the Navy and Marine Corps were also looking for replacements for their respective existing helicopter platforms through efforts unrelated to the UTTAS program. The Navy needed a multipurpose platform that could be sea-based and could use the Light Airborne Multi-Purpose System (LAMPS), an avionics system designed primarily for antisubmarine warfare. The newest LAMPS (Mark III) was being designed at the time but could not be integrated into the existing Navy helicopter platform, the H-2, because of significant equipment size and weight differences between the new and legacy versions of LAMPS. Based on the possibility of savings, the Navy decided to leverage the ongoing H-60 airframe design effort as the platform to field its next-generation avionics system in 1974 and to make changes as necessary to accommodate Navy-specific needs. This included adding Navy-specific tests to the battery of Army prototype tests and exploring airframe modifications to make the H-60 a viable platform for use at sea (e.g., accommodating berthing in small surface ship aircraft hangars).[56]

It is important to note that the Navy did not explicitly tie its program's design down-select to the Army's down-select. The Navy independently submitted an RFP to which the UTTAS competitors responded and separately selected Sikorsky a few months after the Army had; the new platform was called the SH-60B Seahawk. The Army's selection of Sikorsky was considered a positive factor in selecting the H-60 design but was not the only reason the Navy selected Sikorsky's design. According to two GAO reports, one on UTTAS and another on LAMPS, other major factors that led to each service's decision to select Sikorsky included a relatively low bid, significant prior military helicopter program experience, and comparable performance in the

---

55 GAO, 1977.

56 GAO, *The Light Airborne Multi-Purpose System, LAMPS MK III, Progress Evident but Some Problems and Questions Remain*, Washington, D.C., C-MASAD-81-4, February 23, 1981.

test program.[57] As for the Marine Corps, while it considered the H-60 for use as a Huey replacement for troop transport, ultimately it chose to continue using CH-53 and UH-1 helicopters because of differences in Army and Marine Corps squad size and distance requirements.

The Sikorsky H-60 design has had a long service life for the U.S. military, and multiple variants are still in production and use. The first Army and Navy variants of the H-60 were known as the UH-60A and SH-60B and entered service in 1979 and 1984, respectively. Since that point, about 20 variants have seen service time for different branches of the U.S. military, including for the Coast Guard. Currently, the Army uses the UH-60M for different types of operations (e.g., utility transport, medevac); the Navy uses the MH-60S for different purposes (e.g., multimission utility transport, special operations), as well as the SH-60R (specifically for antisurface and anti-submarine warfare). Each variant carries mission-specific equipment and is modified as necessary to support service-specific missions. Additionally, Sikorsky has sold versions of the H-60 to U.S.-allied foreign militaries under the designator S-70 for decades.[58]

The H-60 helicopter was not a joint program. The program was established and operated as a single-service program; other services' involvement in the program was voluntary as opposed to having been mandated by a higher authority. The Army did not conceive of the UH-60 as a joint program but, rather, as a program to fulfill the service's specific needs and requirements. In need of a new platform shortly after the inception of the UH-60, the Navy capitalized on Army developments that were under way. While both the Navy and Army executed service-specific tests as part of the same prototype test program, each interested party (including the Marines, who ultimately did not get involved in the design phase) maintained its own set of operational requirements, put a high priority on maintaining these requirements, and selected an airframe that best suited its needs (although having a similar airframe was considered a plus in the down-select). The ele-

---

[57] GAO, 1977 and 1981.

[58] IHS Jane's, "Sikorsky S-70B," *Jane's All the World's Aircraft*, August 12, 2015.

ments of commonality across the H-60 variants were not mandated but, rather, emerged organically.

The similarities between the two airframes have allowed for multiyear contracting strategies in which the two services purchase multiple years of airframes for multiple variants together in bulk; these multiservice, multiyear contracts have been used eight times so far on the H-60 program. The Navy has had a second multiyear contract with the supplier for its service-specific mission equipment.

According to a former Navy SH-60B LAMPS pilot, the H-60 is not considered a joint program, and its relative success across so many different services and mission areas is due to the qualities of its basic airframe and a modular approach to mission system equipment:

> The MH-60S and MH-60R are both highly modular, and their mission sets [and] capabilities are mostly defined by how they are configured at the moment. Between the two, they have a wide mission set. The capability sets are less about the airframe, and more about what modular systems can be integrated into the airframe. The success of the H-60 airframe over the years has a lot to do with starting with a very solid, reliable, forgiving (if somewhat unspectacular) airframe and taking the modular approach to mission capabilities.[59]

One derivative form of commonality relevant here, within the Army, is the engine. The UH-60 and AH-64 Apache share the same engine, although they are different aircraft with different missions and operational profiles. But this does illustrate that major subsystem commonality across platforms with different missions and capability sets is possible.

In terms of program outcomes, cost and schedule changes during the program design and prototype phases were relatively minor. Certain individual contract milestone dates shifted by a few months, and overall program costs came in below the expected value a few years after the down-select to Sikorsky ($1.942 billion in FY 1971 and $1.704 billion in FY 1979—both in FY 1971 dollars), although cost did not uni-

---

59 RAND team discussion with former Navy SH-60B pilot, January 2016.

formly decrease over the program's life. Program cost slightly increased during the design phase (seemingly because of the presence of challenging requirements that the program office was unwilling to relax) but ultimately decreased during the procurement phase.

Later SARs for subsequent H-60 variants (UH-60L, SH-60F) show some cost growth. Cost growth on the Army's side of the H-60 program in the 1980s and 1990s was related to increases in the number of ordered aircraft (1327 to 2257). Cost growth on the Navy's side during this time frame, however, was related to a new sonar system, increased spare quantities, and additional ordered support items (among a few other smaller factors). In summary, while cost growth existed in the multiple parts of the H-60 helicopter program, it did not seem to be related to any joint management activity or elements of commonality among the system variants but, rather, was due to the services' needs for their respective variants of the H-60.

The H-60 system variants emerged from service-specific needs, with minimal coordination of requirements development and joint activities. The Army clearly was the lead service for the platform and engine, with the Navy and eventually the Air Force and SOCOM funding their own mission system development. The program reflects elements of both a lead service and multiservice model, as well as demonstrating how some degree of commonality may emerge organically without joint management constructs.

We were unable to locate an analysis of the benefits (i.e., cost savings or interoperability) of the commonality present among the H-60 variants.

Lessons from the RAND Joint Fighter Aircraft Study

One of the few systematic analyses of joint program management was conducted by a RAND team and published in two volumes in 2013. Sponsored by the Air Force, the objective of the study was to determine whether the theoretical cost savings presumed attributable to a joint future fighter aircraft program were large enough to mitigate the cost premium associated with joint program management.[60]

Joint program management theoretically generates cost savings by reducing duplication in RDT&E activities (i.e., multiple services executing technology or development programs against the same or similar mission requirements) and by using economies of scale in procurement (increased quantities and a longer cost-improvement [learning] curve) and operations and support (i.e., shared supply chains, depot support). However, joint program management tends to increase both technical and programmatic complexities, which then drive up costs. This increased complexity is related to the need to integrate the different sets of requirements of the participating services while maximizing commonality to maximize cost savings. Another set of costs associated with commonality is potential increased operating costs or reductions in performance caused by accommodating different requirements in the same design.

The researchers found that design and system commonalities are the basis for the theoretical joint cost savings.[61] Using available data on historical joint aircraft costs and cost growth, the research also found that even under assumptions that clearly benefit the joint program (i.e., 100-percent commonality is achieved), there are no scenarios in which the cost savings of a joint fighter aircraft program are higher than the cost premium and higher cost growth attributable to joint management. This analysis applies to both the general case and to the specific case of the JSF experience.[62]

The lesson here for FVL is clear: Joint program management of a large, complex program like any of the expected FVL FoS programs is not likely to result in cost savings unless the participating services have identical requirements. The report also notes the potential increased risk to the industrial base (joint programs necessarily reduce the number of new programs, leading to industry consolidation and a reduction in

---

[60] Lorell et al., 2013.

[61] This is similar to our finding that the benefits of joint management are attributable to commonality.

[62] Lorell, 2013, pp. 11–17 and p. 32.

the breadth and scope of design and development experience) and to operations (e.g., the potential grounding of the entire fleet because of a design flaw; less responsiveness to emergent adversarial capabilities).

Other Historical Analyses

Many of challenges associated with joint program management have been known and documented for many years. Two older reports on DoD's joint program management experience offer insights still relevant today. In particular, these older reports identify the same set of factors affecting joint program management as recent research does and also identify elements of joint program management that help enable successful program execution.

A 1983 GAO report observed that developing joint requirements is the main challenge to overcome. There is a natural resistance to joint requirements among the services, which have had difficulty reaching agreement. Program mergers are often directed or arranged too late in the process (i.e., after single-service requirements are developed). GAO also noted that rigid specifications can be problematic. The rigidity can be due to culture, already validated requirements, or programs are too far along when merged. The services also have different organizational arrangements, standards, data requirements, manuals, provisioning, integration of military specifications and standards, occupational skills, training methods, and test requirements. Logistics processes were noted as being especially difficult to make common. Nomenclature and differences in interpretation of both policy and acquisition concepts can further complicate things.[63] The lesson appears to be that directed mergers to form a joint program present challenges that are very difficult to overcome. In a lead service structure, the GAO noted issues including program office representation, whose rules, processes, and decisionmakers are used, and budget commitment and control. Finally, GAO noted that there is generally no penalty incurred for pull-

---

[63] GAO, *Joint Major System Acquisition by the Military Services: An Elusive Strategy*, Washington, D.C., NSIAD-84-22, December 1983, pp. 15–16.

ing out of a joint program. On a more positive note, GAO suggested using the joint PM in an "orchestrator" role[64] and that voluntary participation is important.

A 1983 DSB study had similar findings. DSB associated problems with joint management of major programs with lack of convergence on requirements and doctrine, shifting service priorities, funding instability, changes in perceived threat, and an ad hoc management environment.[65] Program failures stemmed from little attention early in the planning process to requirements and management issues. The ad hoc management environment is due, at least in part, to these factors, as well as the lack of guidance on joint program management found in DoDI 5000.2 in effect at that time.

DSB noted that nonmajor programs and S&T activities tended to have more success than MDAPs. The reason these were more successful is that smaller programs are more manageable.

DSB's recommended management construct was a formal institutionalized process in which the joint staff and the services ensure that the prerequisites for program success are present prior to program initiation. These prerequisites include establishing a firm foundation, in terms of both requirements and management processes. This includes evaluation of technologies, requirements, program options, and a dispute resolution process. DSB also recommended using existing organizations and processes to manage joint programs, rather than creating all-new organizations. Procedures should be formally captured in DoD's governing acquisition policies (e.g., DoD 5000 series). DSB also recommended that joint programs should be funded by a single service that acts as the lead agent, with service-unique items funded directly by other participating services. An MoU that documents agreements is an important part of the joint program's foundation. Technical information-sharing across services regarding results from S&T and R&D activities is also a key enabler. The lead service–executive agent construct was identified as relatively better than other management constructs in part because it

---

[64] GAO, 1983, p. 26.

[65] Defense Science Board, *Summer Study Briefing Report for Joint Service Acquisition Programs*, August 1983.

limited the number of decisionmakers. Using established organizations and roles, responsibilities, and authorities was also a key enabler.

Interestingly, the DSB study noted that logistics savings resulting from commonality were possible only if data, publications, test equipment, training, and spares were truly common or identical.

Lessons for FVL

There are several important lessons for FVL from this review of historical DoD experience with joint program management. Here, we summarize the key lessons for FVL from each program reviewed.

JIAMDO represents an approach to coordinating requirements, operational concepts, and equipment across the services over time. It has a budget and a full-time staff, which allows it to sponsor demonstrations and conduct analyses to help persuade services to incorporate common concepts, requirements, and equipment in their missile defense programs. For FVL, this kind of organization illustrates one approach to providing an important long-term coordination function.

JLTV was a joint management construct from inception, with the Army designated as lead service, including use of Army acquisition and approval processes, except for a select few milestone documents that the Marine Corps also approved. The Marine Corps was represented in program leadership and also provided significant RDT&E funding. Procedures and responsibilities for joint management were defined and documented up front in program planning. Requirements were largely the same; in the few instances in which they were not, the services worked to accommodate the concerns of the other service. JLTV represents a successful joint management construct, based on a lead service model with significant joint decisionmaking and funding. This model may be appropriate for FVL programs in which the basic requirements and performance specifications are similar.

JPALS' original joint management construct is an interesting model for FVL. Recognizing that the services had the need for the same basic capability but with somewhat different requirements and timing, decisionmakers divided the program into capability-based increments in which the lead service for each increment was the service that needed that specific increment of capability. The level of joint participation was expected to change over time. It is not known how well this would have worked, because the program was restructured in response to changes in need, but it does represent an approach potentially appropriate for the FVL FoS (i.e., lead service for each capability set or program) or for select major subsystems required across FVL programs.

JTRS is a case in which the complicated joint management construct clearly had a negative effect on program execution. The initial joint construct included a weak lead service model, with the service managing the radio set most related to its need. Part of the issue was the lack of a decisionmaking process or other mechanism that could quickly address conflicts in requirements. The program was restructured to a JPEO and joint program construct under Navy leadership. The program was also designated as an acquisition reform pilot program that allowed it to define its own decision processes. However, urgent operational needs resulted in the services procuring non-JTRS radios for their deployed forces. This new structure was dissolved after a few years; it is not known whether it would have resolved the joint management issues associated with the program had it continued. For FVL, JTRS illustrates a less-than-successful JPEO–JPO management construct. However, JTRS also illustrates how some commonality and interoperability can be accomplished through software rather than hardware.

JSTARS is an example in which existing service programs were merged to form a joint program using the lead service model. The program used an MoA to alleviate services' concerns. JSTARS is unusual in that the specific aspect of the system eventually operated by the participating services was different (airborne sensor versus ground-based data processing). As such, it illustrates services working together to develop an operational concept that satisfies both services' mission needs.

The V-22 program provides an example of at least two issues that should be of importance to FVL stakeholders. First, the degree of joint participation may change during program execution as a result of changing priorities or circumstances in individual services. Second, joint program management can increase the technical complexity of a program as the services' different requirements are accommodated. This increase in technical complexity is in addition to the increase in management complexity entailed in multiservice endeavors. That said, the V-22 also represents a model of joint management in which a basic system design forms the basis for variants for other services. This type of joint management requires multiservice participation up front to reach agreement on the basic variant, with the services expecting to fund their own mission-specific modifications and equipment. The use of an MoU capturing this agreement early in the program is notable.

JDAM merged separate Navy and Air Force programs in 1991, with the Air Force designated as lead service. The Defense acquisition pilot program (DAPP) designation and related acquisition reform initiatives, stable requirements, enhanced PM authority and autonomy, and senior leader support enabled a positive outcome. Problems and issues associated with joint management appear to have been mitigated by services having the same requirements and acquisition reform efforts; joint management did not play a significant role in outcomes. For FVL, the JDAM experience suggests that merged programs can work if requirements are the same and that the lead service construct can work well under the right conditions, including stable requirements and senior leader support.

JPATS began and has operated as a fully joint program since its inception. The platform commonality agreed on at the start of the program and the willingness to compromise on certain technical decisions enabled and sustained the successful joint management effort. Additionally, agreement on system requirements allowed both services to purchase the same final product on one contract, possibly also leading to cost efficiencies. JPATS represents a way for participating services to coordinate their requirements capability inputs during program execution. Pilot training is driven by service culture and mission requirements. This program illustrates that different services can potentially collaborate on nonmateriel aspects of an acquisition. While the root cause analysis documented a series of technical problems experienced during program execution, joint management was not identified as a significant factor.[66]

The H-60 system variants emerged from service-specific needs, with minimal coordination of requirements development and joint activities. The Army clearly was the lead service for the platform and engine, with the Navy and eventually the Air Force and SOCOM funding their own mission system development. The program reflects elements of both a lead service and multiservice model, as well as demonstrating how some degree of commonality may emerge naturally without joint management constructs.

---

[66] Arena et al., 2014.

CHAPTER SIX

Alternative Management Constructs

To make a comparison of alternative management constructs meaningful, we made two specific assumptions about what is being managed:

- **The FVL FoS:** A series of distinct programs, each of which is rolled out as requirements mature and are validated, needs or capability gaps are identified, and funding becomes available.
- **Commonality:** Enable, facilitate, and support commonality within a program and across programs. This includes a significant time component as a specific program matures and moves through its life cycle, as well as evolving to meet new threats or take advantage of emerging technologies.

Each program reflects a specific air vehicle type and weight class, corresponding to a specific capability set and associated performance specifications. Each program addresses, in whole or in part, the requirements and needs of at least two services. Common elements can include both hardware and software and can be associated with air vehicle, engine, or mission systems. The time frame for the overall FVL initiative is long term (decades), with multiple distinct programs and the need for adaptation in FVL organizations. There will be FVL-related activities in multiple phases of the life cycle—requirements development (ICD and CDD, CONOPS, joint AoA), S&T activities (e.g., technology demonstrations), program (development and procurement), and sustainment (operations and maintenance)—and these phases will have overlapping time frames.

Management constructs will need to manage the FVL FoS over time and set the conditions for some degree of commonality to be achieved and maintained.

These assumptions are derived from and are consistent with the *FVL Strategic Plan*, other official program documentation, and presentations and discussions we observed at JCoC meetings over the period from May 2015 to February 2016.

There is a wide range of management constructs possible for FVL. It is not possible to adequately describe the advantages and disadvantages of all of them. In addition, the specific advantages and disadvantages (benefits and costs) of a particular management construct are contingent on a clear definition of exactly what is being acquired by whom in what future business, organizational, and budgetary environment. Rather, we have chosen to describe two idealized structures at the extremes of the joint management spectrum and use those constructs as a way to discuss variants that could be applicable under different circumstances. This is not intended as an exhaustive list but rather to enable illustration of the issues that the FVL stakeholders should consider as they contemplate the joint (and common) aspects of the FVL initiative going forward.

The analysis of management constructs presented in this chapter draws on the literature review and historical experience discussed earlier. The review of organizational theory and business management literature in Chapter Four helped us identify the dimensions or key characteristics that define a management construct and how these characteristics may vary across them. The lessons from DoD's historical experience discussed in Chapter Five were used to help identify the advantages and disadvantages (costs and benefits) of different constructs.

The design of a management construct—an organizational structure and its related business processes—can be described by a basic set of parameters and functions. In general form, these include

- size of organization, including workforce size and number of business units or components

- structural elements, including hierarchy (number of layers), the distribution of authority (vertical and horizontal control), the degree of centralization, and grouping (divisions, departments, functional, geographical, product, multiproduct, virtual)
- processes and mechanisms for communication, coordination, and integration, including both formal and informal mechanisms within the established hierarchy
- location of decisionmaking relative to sources of information; in general, decisionmaking closer to where the work is being performed and the information generated supports more agile decisionmaking and a more adaptive organization.
- workforce management, including tenure and turnover patterns, skill sets, and the degree of professionalism required.

These features or descriptors of a management construct are not unique to DoD. Rather, they are generally important and applicable when describing an organizational design and business or process rules in both the public and private sectors. Being on one end of the spectrum or the other on any of these parameters (e.g., vertical [hierarchical] versus horizontal [flat] structure, centralized versus decentralized decisionmaking) is neither good nor bad in itself. The key to program success is understanding the combination of these features that best addresses a particular situation.

Joint Structures

Figure 6.1 is a generalized template intended to maximize commonality through joint management of the FVL FoS. At this general level, it comports to existing policy and regulation, which provide for JPEOs and JPOs. When we overlay the existing FVL initiative management construct on this template in Figure 6.1, we place the ESG, JCoC, JCIC, and IPTs on the requirements side because most of the work to date has concerned development of joint requirements. In this more-permanent structure, the existing organizations remain co-chaired by OUSD(AT&L) and the Joint Staff, but they become more institution-

Figure 6.1
Joint FVL Management Construct Template

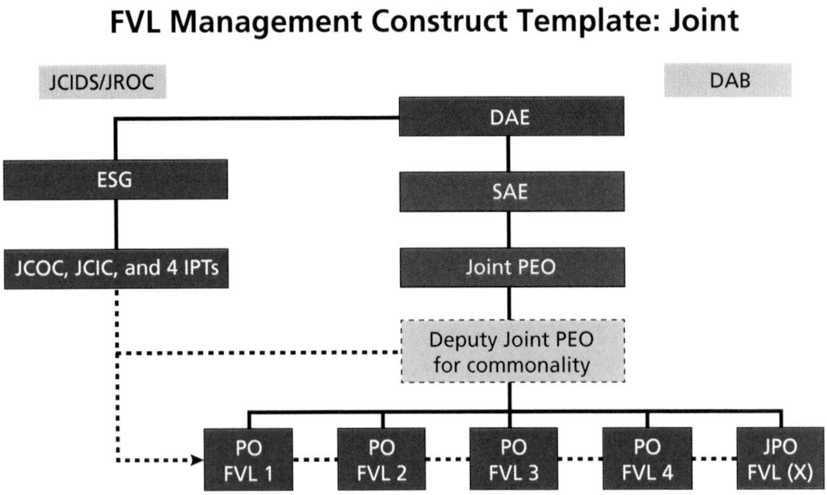

NOTE: SAE = Service Acquisition Executive.
RAND *RR2010-6.1*

alized in the JCIDS and acquisition oversight processes. The rationale for this is that the FVL FoS will require management and coordination over several decades; therefore, roles, responsibilities, and authorities should be institutionalized to provide that longevity.

This structure has two fundamentally different variants regarding the degree of authority invested in the joint organizations. On the requirements side, the JCoC through the requirements IPT could be given the authority to direct and then maintain a stable set of performance specifications for each FVL program. It would also need the authority to direct joint requirements across those programs and over time, since the programs do not all happen at once. Alternatively, the structure could stay largely as it currently is, focusing on coordination and consensus-building instead of mandating specific behaviors or actions.

There are a number of additional variants that emphasize joint management of all or parts of FVL:

A JPO reporting directly to the DAE. In theory, the Service Acquisition Executive (SAE) and associated staff is duplicated at the OUSD(AT&L) level. Oversight could be performed directly by OUSD(AT&L) staff functions. This structure might be best suited to a single, large program responsible for developing multiple variants (i.e., FVL capability sets).

A JPO under single-service PEO, with individual offices for each FVL program (*$FVL_1 - FVL_X$*). In this structure, the FVL programs, still defined by capability sets, are embedded within a large JPO. This structure would also be suitable for a single, large program and might enable improved coordination and collaboration among program-level acquisition organizations but would have a greater challenge integrating with the technology and acquisition organizations of the individual services.

A JPO under single-service PEO, with individual offices by major subsystems (platform, engine, mission systems). Rather than organizing by product-oriented programs defined by capability sets, this structure would organization joint efforts by major subsystem. Thus, a joint platform office would focus only on developing and maturing platform-related design elements that could be used by multiple joint or single-service programs focused on fielding a specific system or capability. This would be a radical departure from how current programs are oriented and would require major changes in how requirements and technology organizations collaborate with acquisition programs, as well as how programs are designed. System integration would take an even more visible role than in current practice.

Another set of variants concerns the more-direct management of commonality. Figure 6.1 also shows a deputy PEO for commonality in the acquisition chain, which would be responsible for maximizing common elements within individual programs and across programs over time. As part of the acquisition chain, a deputy PEO for commonality would focus on ensuring that technology or subsystems and components are used by multiple FVL programs and variants within those programs. Similarly, this position could be a Director for Commonality and Interoperability. A Director for Commonality and Interoperability would lead an acquisition program responsible for developing and maturing subsystems and components that could be used by multiple FVL programs. Responsibility for commonality could also be located within the current ESG-JCoC structure, either in the current CS IPT or as a newly established Director of Commonality and Interoperability. This structure would emphasize common requirements development, or even DOTmLPF commonality that would lead to increased interoperability.

Another option, focused on requirements development, would be a Chairman's Controlled Activity, similar to JIAMDO. This organization would try to ensure that requirements for individual FVL programs are as common as possible. Two variants exist here depending on the degree of authority—whether the organization could mandate the incorporation of joint requirements and common elements (enforced through the JCIDS process and JROC decision) or whether analysis and demonstration is used to persuade programs to voluntarily incorporate joint requirements and commonality. In either case, one would expect this construct to better enable the development and funding of joint requirements and capability integration across programs through improved information-sharing and analyses emphasizing the net benefits of commonality.

Key issues for any of these joint management constructs include managing requirements to maximize joint management; enabling information- and technology-sharing across programs; the relative authority of the joint requirements and acquisition organizations to direct compliance and enforce joint management; and determining and maintaining a decision, approval, and reporting chain that is not overly duplicative across services but still provides stakeholders with needed transparency and the ability to shape decisions that significantly affect their equities. These are some of the key issues that need to be resolved to facilitate successful joint management of FVL.

Joint management constructs obviously and clearly place the emphasis on joint management; these constructs face all the challenges of joint management discussed previously. Some variants described here would require the establishment of entirely new constructs that have no direct precedent in DoD experience. These include the JPEO and deputy JPEO for commonality configurations, as well as JPOs designed around major subsystems and platforms rather than complete weapon systems. These organizations would face the additional challenge of overcoming the resistance of existing organizations operating in the same space. Historical experience (i.e., JSF, JTRS, JLTV, JDAM) suggests that the conditions under which joint management constructs can be successful are very narrowly defined. The most important condition is that requirements are sufficiently similar across all participating services that they can be easily accommodated in system design without also introducing technical complexity or performance trade-offs.

Lead Service or Executive Agent

In a lead service or executive agent model, a single service is designated as the lead organization with the majority of responsibility for planning and managing a program. The lead service model has many possible variations based on the exact responsibilities of the lead service and how the other participating services are included.

A highly generalized lead service configuration that overlays the current FVL management construct and expectations of the *FVL Strategic Plan* is shown in Figure 6.2. In this generalized model, the lead

Figure 6.2
Lead Service Construct

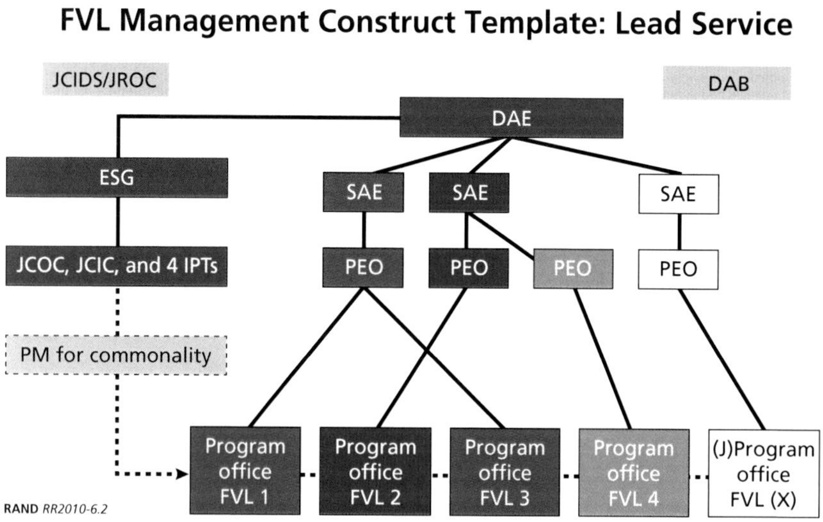

service for each FVL program, still defined as *distinct capability sets*, is determined at program start. That designation can be made based on a number of different criteria, including relative importance of the capability set to mission needs for each service or relative budget contribution to development activities or procurement quantities. Programs would report through their established service PEO-SAE chain.

We have included the existing joint FVL management construct here to enable some degree of continuing coordination and collaboration in requirements development and planning. Without some formal mechanism for coordination among stakeholders, developing and managing an FoS are not possible. However, the roles that these organizations play would be more limited than in the joint model and perhaps reduced from current practice.

There are multiple variants of lead service constructs. The basic variants include the following:

A series of single-service FVL-related programs with no formal coordination or collaboration in acquisition. Minimal coordination of requirements to enable the FVL FoS remains with the ESG-JCoC-JCIC structure; continued stakeholder participation in that structure is entirely voluntary. This construct is at the single-service extreme on the joint and common spectrums. It is suitable when requirements (capability sets) are fundamentally dissimilar across services such that only small components, technologies, or DOTmLPF activities can usefully be common across systems. We would expect some commonality to materialize organically, but it will not be clear how much until FVL is further into system design and development. However, this structure does not preclude the leader-follower model of derivative commonality illustrated by UH-60 over time.

A lead service program with formal program office representation from other participating services. The major difference between this construct and a single-service construct is that the program office would include personnel from other services in key leadership positions. For instance, the PM would be from the lead service, while the deputy PM would be from a participating service. The heads of the functional branches (e.g., engineering, test, logistics, acquisition) would similarly include personnel from the lead service in the lead position, with personnel from the other participating services filling the deputy positions. Decision, approval, and reporting chains would remain with the lead service, although participating services could be included for select areas. Participation and rules of engagement could either be informal or codified in an MoA to ensure consistency. This construct could better enable commonality through the direct participation of stakeholders at the program office level. It would also improve communication and information-sharing among participants above what would be expected in a single-service model. The main difference between this construct and a JPO is that management, decision-making, oversight, and approval chains rely more heavily on lead service processes and organizations. Those functions are more shared and often joint in the classic JPO structure.

A series of lead service programs with representation from other services or single-service programs with specific components or subsystems developed and matured by other services is another possible variant. The programs contributing subsystems and components could be managed outside the system-level program office.

Lead service common subsystem programs. The FVL FoS could be structured around a set of subsystem development and maturation programs, each of which could be structured as either single service or lead service with representation. This is essentially a construct implementing parallel development (intentional separation of management and technology development of subsystems from each other instead of a single program developing and managing air vehicle, engines, and mission systems) and would represent a departure from traditional acquisition organizations.

Lead service management constructs have to address similar challenges as joint constructs, though the degree to which these are challenges may be less. Managing joint requirements and information- and technology-sharing practices is still a challenge for any lead service construct that includes joint participation at the program level. Authority to direct compliance should be largely resolved in the lead service construct, except when other participating services resist decisions made by the lead service that may affect their equities.

One significant disadvantage of a lead service construct is the added difficulty of ensuring coordination, information-sharing, and collaboration across acquisition organizations (programs, PEOs, technology development centers, or labs). Reporting chains for these service-lead programs only come together at the DAE level. This challenge can be mitigated somewhat via MoAs or MoUs established at the outset of each program that include specific provisions addressing these issues.

An additional set of variants is suggested by the VLC, an industry group representing the business interest in future rotorcraft. The VLC has stated that any FVL organizational structure and acquisition strategy has implications for industry structure, investment, and innovation. In a paper prepared for the FVL initiative, the VLC recommended a multiple service (joint) "program of programs" at the PEO level or higher to develop ideas for commonality and ensure implementation. A series of smaller product-oriented programs would each address a specific capability set.[1] In the terms we are using to describe constructs, the program of programs manages the FVL FoS, while each weapon system program is specific to a capability set and the services whose equities are embedded in that capability set. Only the higher-level program of programs needs to be joint; the smaller programs could be joint, lead service with representation, or single service. Derivative commonality is still feasible. Thus, the VLC ideas introduce the notion of hybrid constructs: Some aspects are joint, while others can be lead- or single-service constructs.

Industry, as reflected in the VLC papers, is less interested in the management construct per se and more interested in ensuring that the chosen construct facilitates opportunities for competition that would drive innovation and provide potential business opportunities. Industry would like to see the acquisition time frame be shorter than is currently projected.[2] Ultimately, industry is seeking a more-transparent planning process, as greater information and insights into FVL initiative could help justify investments and could give greater sense of potential returns on investments.

---

[1] VLC, 2015a.

[2] VLC, 2015a.

CHAPTER SEVEN

Conclusions and Recommendations

Overarching Findings and Observations

FVL represents a valuable opportunity to fully explore joint management constructs and commonality. However, it is critical that all stakeholders share the same interpretation of joint versus commonality and recognize the perceived versus real benefits of each. It is important to distinguish between the two concepts. In DoD context, joint program management means two or more services are working together to achieve a shared objective. In that sense, it is both an operational and management construct. Commonality is a process, or a technical or engineering property of a physical (hardware and software) item—part, component, subsystem, or system—that is used in multiple weapon systems or by multiple services. In the acquisition context, many of the benefits expected from joint management come from the commonality that it tries to impart. However, commonality can also emerge organically, without joint management. The key to successful program execution is removing the structural and political impediments that keep organizations from accessing and drawing on the benefits of commonality.

We found that the benefit-cost framework is a useful overarching approach to thinking about whether a particular program should be jointly managed. Joint management adds complexity and additional challenges to the already challenging task of managing large, complex weapon system programs. The joint cost premium can be high for large, complex programs.[1] On the other side of the equation, evidence of the benefits of commonality is sparse. Potential benefits may accrue throughout the program life cycle, including, for example, reduced duplication of effort in development, economies of scale in production, reduced costs for primary training of operators and sustainers, reduced logistics footprint, and greater interoperability among the elements of a joint task force. Some of these potential benefits can be quantified or monetized, but others cannot. For FVL, the net benefit calculation that would inform a decision of how and what should be joint has considerable uncertainty on both sides of the equation. A broad view of potential materiel and nonmateriel costs and benefits of joint management and commonality is needed to fully understand the implications for FVL LCCs.

Joint programs do not necessarily lead to a net positive outcome even if a high degree of commonality is achieved. Danzig argues that the unpredictability of future events requires changes in both the designs of systems and the processes used to acquire them, with an emphasis on both adaptability and agility.[2] As part of that argument, he notes that competition among and differences between the services inherently provide agility with respect to unanticipated threats. If the services all used the same weapon system (e.g., maximizing commonality), then the risk associated with technical deficiencies in the system as well as changes to the threat may be higher. Lorell and his colleagues also make this argument with respect to joint fighters.[3] This risk should be included when evaluating the extent to which joint management or commonality should be emphasized in FVL.

Joint management of smaller programs, subsystems, and components is more likely to yield benefits.[4] Smaller programs are less complex, more focused, and more manageable. This suggests that a management construct for managing specific elements of the avionics suite,

---

[1] Lorell et al., 2013.

[2] Richard Danzig, *Driving in the Dark: Ten Propositions About Prediction and National Security*, Washington, D.C.: Center for a New American Security, October 2011.

[3] Lorell et al., 2013.

[4] DSB, 1983.

mission systems, or engines could potentially be managed under a joint construct, as long as the services' requirements for these subsystems are the same.

Multiple management constructs are possible; there is no single definitive answer to the question of which construct would be most suitable for FVL. In fact, different constructs may be required for each FVL program. Adaptation over time and responsiveness to change are important attributes for any FVL management construct. However, the management construct that appears most successful in fielding useful systems with some degree of commonality is a lead service construct on a basic platform with each participating service funding and managing its own mission systems. The V-22 and H-60 exemplify the elements of this particular construct. The CH-47F/MH-47G also fits this model; SOCOM modified an existing airframe with its own mission equipment. A few joint programs—JLTV, JDAM—suggest that a joint management construct can be successful but only under very specific and narrowly defined conditions, including that accommodating the differing requirements of participating services does not introduce technical complexity or risk or require the services to accept significant performance shortfalls.

There is a semantic debate over whether these examples are truly joint programs; we would argue that they are not joint from the start but rather use a form of derivative commonality. In any case, the programs illustrate a multiservice approach to managing commonality that may be less complex than true joint program management.

Attributes and Enablers of Successful Joint Efforts

Prerequisites for joint management success can be identified, but they alone may not be sufficient to enable success. DoD's historical record of joint program management and findings from the business management and private-sector joint-venture literature offer lessons for any future joint management effort. Although we can identify the important attributes of joint management constructs and enablers of successful joint efforts, implementation remains a significant challenge. These concepts are necessary, but not sufficient, conditions to promoting successful joint efforts.

Ensure that mission requirements are substantially the same. This condition ensures a strategic alignment of interests and equities among the participants in a given program or series of programs. As is the case with MDAPs in general, requirements—joint or otherwise—must be demonstrably feasible with current technology. Taking a more-conservative approach to design (i.e., working with more-mature technologies, settling on a stable set of requirements, and taking a more-incremental approach to development and acquisition) could lead to better outcomes.

Ensure voluntary participation. This attribute means that, to a large extent, participation in a joint activity is both self-motivated and self-organizing. It means that each participating service has thought through the benefits and costs of joint participation or management to achieve increased commonality, and it has concluded that the arrangement will result in net benefits in the long term.

Perform detailed planning up front. Identify the feasibility and cost-performance of commonality, broadly defined, and reach agreement on exactly which aspects of system development, operation, and sustainment should be common for each program or subprogram.[5] Roles, responsibilities, and authorities need to be established early in planning, including both routine approval and decision processes, as well as adjudication processes in the event of a conflict among participants. Processes to manage change in threats, technology, requirements, institutional environment, funding, and improved information also need to be established. An S&T road map intended to feed new concepts and technologies into future programs can be part of this planning function.

Codify the plan. A formal MoU or MoA should be used to codify the agreements reached during early program planning. At a

---

[5] There seems to be agreement that early system engineering activities can facilitate this for joint programs. For discussion of early SE in single-service programs, see GAO, *Defense Acquisition Process: Military Service Chiefs' Concerns Reflect Need to Better Define Requirements Before Programs Start*, Washington, D.C., GAO-15-469, June 11, 2015.

minimum, this should include explicitly laying out roles, responsibilities, and authorities; processes to resolve conflicts; and processes to adapt to change. The formal agreement should be in place prior to program start, and it should be signed from the program level up through the senior leadership of the requirements and acquisition functions. It should address both the terms of engagement and the terms of disengagement. Codifying the plan helps to ensure consistent execution over time and somewhat mitigates the problem of turnover in execution and oversight officials at all levels.

Ensure senior leader support. Secure and sustain support from senior leadership for the new initiative. *Senior leadership* as defined in this report includes the requirements and capability development community, as well as the acquisition community. Most importantly, it also includes the warfighter. It may be useful to secure the support and agreement of the service chiefs, as well as one or more COCOMs representing the warfighter. This last ensures that the capabilities of the system truly meet the needs of the warfighter.

Establish cross-unit teams. Cross-unit teams or liaison positions and functions facilitate open communication, information-sharing, and transparent decisionmaking. The governance structure should balance shared decisionmaking and oversight with the need to maintain clear command and control. Liaison mechanisms to encourage mutual adjustment within and among units include specific liaison positions, task forces, standing committees, and integrating managers.

Identify and retain evangelizers. It will be important for the FVL initiative to identify and support those key stakeholders who are well positioned (and willing) to advocate for the FVL vision that is decided on, particularly at the middle management and working levels. In certain cases described in Chapter Five, senior leadership actively backed and gave top cover to the new initiative and, in effect, to its leader as well, thereby strengthening the chances of successful outcomes (e.g., JDAM). This is also an important element in Kotter's change management concept. This also may be helpful in the context of FVL. In addition to high-visibility individuals in positions of authority, uniformed service advocates throughout the FVL hierarchy also may help garner support for the FVL initiative (i.e., advocating through education and demonstration).

Ensure that appropriate technical enablers are in place. Open systems and open architecture designs (interfaces, specifications and standards, size, weight, power, and other factors) are critical to achieving some degree of commonality. In addition, use of common design and engineering tools, modeling and simulation, and studies and analyses can facilitate commonality across the spectrum.

Secure incremental successes. Afford the PM with the autonomy to make decisions and secure early successes. Demonstrating incremental success in elements of the program provides a foundation for enabling and securing future successes.

Create the right environment for innovation. In many of the more successful joint cases we examined, decisionmakers created the right environment for innovation and disruption in management practices, some even pursuing DAPP designation. Creating the right environment for innovation is a twofold effort; the first aspect is really about the environment itself—the rules, the flexibility, the tolerance for risk—whereas the second aspect is about the authority afforded to the individuals who would actually be driving the innovation. Those who are leading FVL should commit to the stability and security of the initiative by ensuring that all those involved in the development and execution have sufficient time and resources to succeed.

Minimize the distance between those with the critical information and those making decisions. It may be worthwhile to explicitly define the internal mapping of the FVL organization. This would include defining the formal reporting relationships within the organization and articulating the authorities of the different liaisons, standing committees, services' respective hierarchies, and those responsible for the coordination and continuity of the initiative. In certain cases we examined (JDAM and JTRS), efforts were made to reduce the distance between top decisionmakers and program leaders, thereby compressing decision hierarchies and facilitating greater information-sharing.

In addition to compressing decision hierarchies and limiting the distance between those with information (including those who will be ultimate operational users) and those who are making decisions, the question of personnel is also important: Who should be hired, should they be military or civilian, and how long should they be retained? The hiring of the right personnel is always significant, but particularly so in the context of new initiatives that may benefit from having influential figures driving change and securing buy-in from stakeholders. As important as the organizational construct may be, success in some contexts may be personality-driven.

Recommendations

We believe that even if FVL becomes a series of independent single-service programs, some degree of commonality will emerge organically. If FVL stakeholders want to increase commonality above this level, then some form of higher-level coordinating body or active management construct is required. This management construct does not need to be a JPEO-JPO construct, but some degree of sustained multiservice participation will be necessary. An organization to facilitate information flow and coordinate S&T investment and outputs, requirements development, and budget planning would be needed. Continuation of the ESG/12-pack, JCoC, IPTs, and JCIC is likely the best option for this, although the frequency of meetings and activities can be adjusted to meet the needs of the participants in different phases of planning, analysis, and program execution.

There is no evidence that joint management produces net benefits for a complex program, let alone a series of complex programs as envisioned in the FVL FoS. Joint programs held to be successful (i.e., that resulted in some expected benefits) tend to be relatively less complex (e.g., JLTV and JDAM), whereas more-complex programs do not demonstrate the expected benefits and, in some cases, may incorporate significantly less commonality than originally expected (e.g., JSF). Therefore, **we recommend that FVL *not* implement a purely joint construct at this time**. However, FVL does represent an opportunity to more fully explore management constructs for commonality, analyze the key actions or decisions that generate commonality, and better understand the costs and benefits associated with commonality.

We recommend that FVL take advantage of this opportunity by including such an analysis as part of the upcoming AoA. Metrics for the costs and benefits of commonality and joint management can be established. Actual data from the UH-60 and V-22 programs can be used to identify and quantify both the costs and benefits of the commonality achieved to date and the management constructs used.

We also recommend that any future FVL management construct use existing organizations, authorities, and processes to the maximum extent possible. This argues for a lead service (executive agent) construct with representation from other participating services as appropriate. **We recommend that the attributes and enablers of successful joint management discussed above should also be followed as much as possible**, meaning that they are designed into the planning and implementation of the management construct. Again, the current FVL construct of ESG, JCoC, IPTs, and JCIC is a good foundation to build on (i.e., voluntary participation, planning and coordination, open lines of communication). The importance of having a clear and explicit understanding of requirements (joint or otherwise), of having detailed and realistic plans in place, and of using formal MoAs to codify agreements (e.g., budgeting, decision authority, roles and responsibilities) cannot be overstated.

For joint, lead service, or single-service programs, requirements management is the key to success. This is much harder in a joint or multiservice setting, as different stakeholders may have conflicting missions and interests (and therefore conflicting requirements). Moreover, the relatively longer timeline for joint or multiservice programs means that there is a higher risk of instability (e.g., in threat, services' preferences, technological superiority, and key personnel). The implication is that the program should move faster to compensate. Ideally, leadership would align requirements, identify the mature technology to achieve those requirements, fully fund the effort, and then execute within a short time frame (all while ensuring that the officials who made key decisions and reached the underlying agreements stay through execution of the program). This, in turn, implies the use of more mature technology, stable requirements, and an incremental/evolutionary approach to fielding capabilities. This approach is consistent with FVL's long timeline and with the plan (as stated in the *FVL Strategic Plan* and elsewhere) to spin out a program or subprogram for each capability set.

Given the complexity of the FVL FoS and the long period of time over which its programs will be active and operational, **we recommend consideration of a parallel development approach to common systems**. Parallel development is not concurrency; rather, it is the intentional separation of both technology development and management of major subsystems or components (i.e., elements of mission systems, engines or airframes for a specific weight class or capability set). Parallel development offers a way to decompose a complex program into more manageable parts and introduce new capabilities incrementally over time. The focus here is on managing a core set of common platform, avionics, engine, or mission system subsystems and components using a parallel development approach. It is important to start small—smaller programs tend to be less complex and more manageable. This common system program can be based on a lead service (executive agent) construct with multiservice participation and should follow the same set of attributes and enablers that the larger FVL system programs should. Careful system engineering planning, open systems and open architecture design approaches, and mature and fully understood technology should be used to mitigate the inherent system integration risk of parallel development.

The decades-long time frame of the FVL FoS also suggests that any **management constructs should incorporate the attributes of learning organizations.**[6] Learning organizations are open systems that adapt and thrive in changing environments. Learning organizations are centered on a horizontal structure (so that the highest-ranked decisionmakers cannot become bottlenecks), an adaptive culture that reduces rigidity, a collaborative strategy as opposed to one that is purely competitive, open lines of communication and information-sharing across boundaries, and an emphasis on empowering roles instead of assigning tasks. These attributes mirror and are complementary to the

---

[6] Mark H. Moore, *Recognizing Public Value*, Cambridge, Mass.: Harvard University Press, 2013.

attributes listed above that enable successful management of commonality. We recommend that any management construct adopted by FVL also include these attributes.

We also recommend that the FVL community identify the key framing assumptions that underlie the FVL FoS concept. Framing assumptions are the often-unstated assumptions underlying an MDAP, which, if not true, will have a significant adverse effect on program cost, schedule, and performance outcomes. They are often made early in the planning process but can also occur later in a program's life cycle, particularly for long programs for which the threat or institutional environment has changed.[7]

FVL is in the early planning stage, and framing assumptions related to joint program management can be identified. Our analysis suggests the following framing assumptions for joint FVL:

- Requirements are sufficiently similar to enable commonality.
- Participants (stakeholders) are sufficiently invested so they will not drop out.
- Rules of the road for joint management can be identified and agreed on prior to initiation of a PoR.
- Active management of commonality will enable net cost savings and other benefits across participants (cost savings will be greater than the costs of joint management).

While these statements may seem obvious, it is important that they be made explicit as the FVL stakeholder community continues planning and analysis activities. At this point, these assumptions cannot be proved to be either true or false, which results in a great deal of uncertainty. The failure of any one of these assumptions would significantly affect the viability of any joint or multiservice FVL management construct. We strongly urge the FVL stakeholders to formally develop a set of framing assumptions and then carefully monitor program (and FoS) execution to detect potential early indications that key assumptions may not be true.

---

7 Mark Arena and Lauren Mayer, *Identifying Acquisition Framing Assumptions Through Structured Deliberation*, Santa Monica, Calif.: RAND Corporation, TL-153-OSD, 2014.

Abbreviations

3DLR	3-D Long-Range Radar
AMDR	Air and Missile Defense Radar
AMRAAM	Advanced Medium-Range Air-to-Air Missile
AoA	Analysis of Alternatives
BCA	Business Case Analysis
C4ISR	command, control, communications, computers, intelligence, surveillance, and reconnaissance
CAPE	Cost Assessment and Program Evaluation
CDD	capability development document
COCOM	combatant command
CONOPS	concept of operations
CSAR	combat search and rescue
CS IPT	Common Systems Integrated Product Team
DAB	Defense Acquisition Board
DAE	Defense Acquisition Executive
DAPP	Defense acquisition pilot program
DoD	U.S. Department of Defense
DoDI	Department of Defense Instruction

DOTmLPF	doctrine, organization, training, materiel, leadership and education, personnel, and facilities
DSB	Defense Science Board
EMD	engineering and manufacturing development
ESG	executive steering group
FAA	Federal Aviation Administration
FACE	Future Airborne Capability Environment
FoS	family of systems
FVL	Future Vertical Lift
FY	fiscal year
GAO	U.S. Government Accountability Office
GMTI	Ground Moving Target Indicator
HOGE	Hover Out of Ground Effect
ICD	initial capability document
IPT	Integrated Product Team
JCA	Joint Common Architecture
JCIC	Joint Coordination and Integration Cell
JCIDS	Joint Capabilities Integration and Development System
JCoC	Joint Council of Colonels
JDAM	Joint Direct Attack Munition
JIAMDO	Joint Integrated Air and Missile Defense Organization
JLTV	Joint Light Tactical Vehicle
JMR	Joint Multi-Role
JMR TD	Joint Multi-Role Technical Demonstrator

JPALS	Joint Precision Approach and Landing System
JPATS	Joint Primary Aircraft Training System
JPEO	joint program executive office
JPL	joint priority list
JPO	joint program office
JROC	Joint Requirements Oversight Council
JSF	F-35 Joint Strike Fighter
JSOW	Joint Standoff Weapon Program
JSTARS	Joint Surveillance and Target Attack Radar System
JTAMDO	Joint Theater Air and Missile Defense Organization
JTRS	Joint Tactical Radio System
JVX	Joint Vertical Launch Vehicle
LAMPS	Light Airborne Multi-Purpose System
LCC	life cycle cost
LPD	amphibious transport dock
LRIP	low-rate initial production
LW&M	Land Warfare and Munitions
MDA	milestone decision authority
MDAP	major defense acquisition program
MDD	Material Development Decision
MoA	memorandum of agreement
MoU	memorandum of understanding
OSD	Office of the Secretary of Defense

OUSD(AT&L)	Office of the Under Secretary of Defense for Acquisition, Technology and Logistics
PEO	program executive office
PM	program manager
PoR	Program of Record
R&D	research and development
RDT&E	research, development, test, and evaluation
RFP	request for proposal
S&T	science and technology
SAE	Service Acquisition Executive
SAR	Selected Acquisition Report
SOCOM	Special Operations Command
TM&RR	technology maturation and risk reduction
U.S.C.	U.S. Code
UTTAS	Utility Tactical Transport Aircraft System
VLC	Vertical Lift Consortium
VTOL	vertical takeoff and landing

Bibliography

AHS International, "Future Vertical Lift," web page, undated. As of May 17, 2017: https://vtol.org/what-we-do/advocacy/future-vertical-lift

AHS International, Future Vertical Lift overview briefing at Forum 71 annual event, Virginia Beach, Va., May 2015.

Arena, Mark V., Irv Blickstein, Abby Doll, Jeffrey A. Drezner, James Kallimani, Jennifer Kavanagh, Daniel F. McCaffrey, Megan McKernan, Charles Nemfakos, Rena Rudavsky, Jerry Sollinger, Daniel Tremblay, and Carolyn Wong, *Management Perspectives Pertaining to Root Cause Analyses of Nunn-McCurdy Breaches: Vol. 4*, Santa Monica, Calif.: RAND Corporation, MG-1171/4-OSD, 2013. As of May 9, 2017: https://www.rand.org/pubs/monographs/MG1171z4.html

Arena, Mark V., Irv Blickstein, Dan Gonzales, Sarah Harting, Jennifer Lamping Lewis, Michael McGee, Megan McKernan, Charles Nemfakos, Jan Osburg, Rena Rudavsky, and Jerry Sollinger, *DoD and Commercial Advanced Waveform Developments and Programs with Multiple Nunn-McCurdy Breaches, Vol. 5*, Santa Monica, Calif.: RAND Corporation, MG-1171/5-OSD, 2014. As of March 30, 2017: https://www.rand.org/pubs/monographs/MG1171z5.html

Arena, Mark V., and Lauren A. Mayer, *Identifying Acquisition Framing Assumptions Through Structured Deliberation*, Santa Monica, Calif.: RAND Corporation, TL-153-OSD, 2014. As of May 9, 2017: https://www.rand.org/pubs/tools/TL153.html

Arnold, Scot A., Ji S. Byun, Harley A. Cloud, Alexander O. Gallo, Matthew W. Gonwa, Bruce R. Harmon, Prashant R. Patel, Colin D. Sullivan, John R. Hiller, and Patricia F. Bronson, *WSARA 2009: Joint Strike Fighter Root Cause Analysis*, Alexandria, Va.: Institute for Defense Analyses, IDA Paper P-4612, June 2010. As of May 9, 2017: http://www.acq.osd.mil/parca/docs/2010-ida-rca-jsf-p-4612.pdf

Bamford, James, David Ernst, and David G. Fubini, "Launching a World-Class Joint Venture," *Harvard Business Review*, February 2004. As of March 30, 2017: https://hbr.org/2004/02/launching-a-world-class-joint-venture

Behn, Robert, "The Adoption of Innovation: The Challenge of Learning to Adapt Tacit Knowledge," in Sanford Borins, ed., *Innovations in Government: Research, Recognition, and Replication*, Cambridge, Mass.: Brookings Institution Press and John F. Kennedy School of Government, Harvard University, 2008.

Bennett, Nathan, and G. James Lemoine, "What VUCA Really Means for You," *Harvard Business Review*, January 1, 2014.

Blickstein, Irv, Michael Boito, Jeffrey A. Drezner, James Dryden, Kenneth Horn, James Kallimani, Martin C. Libicki, Megan McKernan, Roger C. Molander, Charles Nemfakos, Chad J. R. Ohlandt, Caroline R. Milne, Rena Rudavsky, Jerry Sollinger, Katharine Watkins Webb, and Carolyn Wong, *Root Cause Analyses of Nunn-McCurdy Breaches: Vol. 1*—Zumwalt-*Class Destroyer, Joint Strike Fighter, Longbow Apache, and Wideband Global Satellite*, Santa Monica, Calif.: RAND Corporation, MG-1171/1-OSD, 2011. As of May 9, 2017: https://www.rand.org/pubs/monographs/MG1171z1.html

Boeing Company, *Implementing Acquisition Reform: A Case Study on Joint Direct Attack Munitions*, Fort Belvoir, Va.: Defense Systems College Management, July 1998.

Builder, Carl H., *The Army in the Strategic Planning Process: Who Shall Bell the Cat?* Santa Monica, Calif.: RAND Corporation, R-3513-A, 1987. As of June 14, 2015: http://www.rand.org/content/dam/rand/pubs/reports/2006/R3513.pdf

Christensen, Clayton M., Taddy Hall, Karen Dillon, and David S. Duncan, *Competing Against Luck: The Story of Innovation and Customer Choice*, New York: Harper Business, 2016.

Clark, Colin, "Boeing to Pentagon: Be Careful When You Pause IRAD Programs," *Breaking Defense*, June 14, 2015. As of June 23, 2015: http://breakingdefense.com/2015/06/boeing-to-pentagon-be-careful-when-you-pause-irad-programs

Danzig, Richard, *Driving in the Dark: Ten Propositions About Prediction and National Security*, Washington, D.C.: Center for a New American Security, October 2011. As of May 9, 2017: https://www.cnas.org/publications/reports/driving-in-the-dark-ten-propositions-about-prediction-and-national-security%C2%A0

Data Link Solutions, "Data Link Solutions Awarded $478.6 Million Contract for Production of MIDS JTRS Terminals," June 18, 2015. As of May 9, 2017: http://www.datalinksolutions.net/news/2015-06-18.php

Defense Acquisition University, *Joint Program Management Handbook*, 3rd ed., Fort Belvoir, Va.: Defense Acquisition University Press, July 2004.

Defense Science Board, *Summer Study Briefing Report for Joint Service Acquisition Programs*, August 1983.

Defense Standardization Program, "Policy and Guidance," web page, undated. As of May 17, 2017:
http://www.dsp.dla.mil/Policy-Guidance

Defense Standardization Program Office, *Defense Standardization Program Case Study: The Virginia Class Submarine Program*, Fort Belvoir, Va.: U.S. Department of Defense, DSP-CS-15, 2007. As of May 17, 2017:
http://quicksearch.dla.mil/qsDocDetails.aspx?ident_number=275846

Department of Defense Instruction 5000.02, *Operation of the Defense Acquisition System*, Washington, D.C.: Under Secretary of Defense (Acquisition, Technology, and Logistics), January 2015.

Deputy Secretary of Defense, *Future Vertical Lift: A Strategic Plan for United States Department of Defense Vertical Lift Aircraft*, Version 2, Washington, D.C.: U.S. Department of Defense, October 1, 2011.

Deputy Secretary of Defense, *Future Vertical Lift: Report on the Strategic Plan for the United States Department of Defense Vertical Lift Aircraft*, submission to Congress, Washington, D.C.: U.S. Department of Defense, 2012.

Deputy Secretary of Defense, *Future Vertical Lift: Report on the Strategic Plan for the United States Department of Defense Vertical Lift Aircraft*, update to Congress, Washington, D.C.: U.S. Department of Defense, 2014.

DoD—*See* U.S. Department of Defense.

Doubleday, Justin, "Air Force, Navy Team Examining Commonality for Future Strategic Missiles," *Inside Defense: Inside the Air Force*, January 21, 2016. As of January 26, 2016:
http://insidedefense.com/inside-air-force/
air-force-navy-team-examining-commonality-future-strategic-missiles

Dunn, Richard J., Price T. Bingham, and Charles A. Fowler, "Ground Moving Target Indicator Radar and the Transformation of U.S. Warfighting," Northrop Grumman Analysis Center Papers, February 2004. As of May 24, 2017:
http://www.northropgrumman.com/AboutUs/AnalysisCenter/Documents/pdfs/
Ground-Moving-Target-Indicator.pdf

Fallows, James, "Uncle Sam Buys an Airplane," *The Atlantic*, June 2002. As of May 9, 2017:
http://www.theatlantic.com/magazine/archive/2002/06/
uncle-sam-buys-an-airplane/302509

Federation of American Scientists, "Joint Precision Approach and Landing System (JPALS) Precision Approach and Landing Capability (PALC)," 1999. As of September 23, 2015:
http://fas.org/man/dod-101/sys/ac/equip/jpals.htm

Feickert, Andrew, *The Joint Tactical Radio System (JTRS) and the Army's Future Combat System (FCS): Issues for Congress*, Washington, D.C.: Congressional Research Service, RL33161, November 17, 2005. As of May 9, 2017: http://www.dtic.mil/dtic/tr/fulltext/u2/a443737.pdf

Feickert, Andrew, *Joint Light Tactical Vehicle (JLTV): Background and Issues for Congress*, Washington, D.C.: Congressional Research Service, October 28, 2015.

Freedberg, Sydney J., Jr., "Navy Wants to Work with Air Force on New Nukes: VADM Benedict," *Breaking Defense*, June 19, 2015. As of June 24, 2015: http://breakingdefense.com/2015/06/navy-wants-to-work-air-force-on-new-nukes-vadm-benedict

GAO—*See* U.S. Government Accountability Office.

Gertler, Jeremiah, *V-22 Osprey Tilt-Rotor Aircraft: Background and Issues for Congress*, Washington, D.C.: Congressional Research Service, RL31384, March 10, 2011.

Green, Robert S., "Organizational Change for Acquisition Reform: Proposed Core Competencies," research paper for Air Command and Staff College, March 1997.

Gupta, Anil, and Haiyan Wang, "Beat the Odds in Cross-Border Joint Ventures," *Harvard Business Review*, October 9, 2013. As of March 30, 2017: https://hbr.org/2013/10/beat-the-odds-in-cross-border-joint-ventures

Haber, Alex, and Jeff Jeffress, "Pentagon Must Tread Carefully on 'Joint' Weapon Acquisitions," *National Defense Magazine*, June 2015.

Harris, Howard, and Mark Lewis, "Proposed Leadership Structure for Joint Acquisition Programs," *Defense Acquisition Research Journal*, Vol. 19, No. 1, Issue 61, January 2012, pp. 33–52.

Heifetz, Ronald, *Leadership Without Easy Answers*, Cambridge, Mass.: Harvard University Press, 1994.

Held, Thomas, Bruce Newsome, and Matthew W. Lewis, *Commonality in Military Equipment: A Framework to Improve Acquisition Decisions*, Santa Monica, Calif.: RAND Corporation, MG-719-A, 2008. As of May 9, 2017: https://www.rand.org/pubs/monographs/MG719.html

Hubbard, Nancy A., *Conquering Global Markets: Secrets from the World's Most Successful Multinationals*, Hampshire, UK: Palgrave Macmillan, 2013.

Hynes, Michael, Sheila Nataraj Kirby, and Jennifer Sloan, *A Casebook of Alternative Governance Structures and Organizational Forms*, Santa Monica, Calif.: RAND Corporation, MR-1103-OSD, 2000. As of May 9, 2017: https://www.rand.org/pubs/monograph_reports/MR1103.html

IHS Jane's, "Sikorsky S-70B," *Jane's All the World's Aircraft*, August 12, 2015.

Ingols, Cynthia, and Lisa Brem, *Implementing Acquisition Reform: A Case Study on the Joint Direct Attack Munitions*, Fort Belvoir, Va.: Defense Systems Management College, July 1998.

Inkpen, Andrew C., and Adva Dinur, "Knowledge Management Processes and International Joint Ventures," *Organization Science*, Vol. 9, No. 4, July–August 1998, pp. 454–468.

JCoC—*See* Joint Council of Colonels.

Joint Council of Colonels, Briefing, July 23, 2015a.

Joint Council of Colonels, Briefing V.3, October 15, 2015b.

Johnson, Dana L., Gregory H. Hilgenberg, and Liam P. Sarsfield, *Policy Issues and Challenges for Interagency Space Systems Acquisition*, Santa Monica, Calif.: RAND Corporation, MR-1372-NRO, 2001. As of May 9, 2017:
https://www.rand.org/pubs/monograph_reports/MR1372.html

JPALS—*See* Joint Precision Approach and Landing System.

Joint Precision Approach and Landing System Operational Capabilities Integrated Product Team, *Joint Precision Approach and Landing System (JPALS) Updated Analysis of Alternatives Study Report*, November 17, 2005.

Kavanagh, Jennifer, Megan McKernan, Kathryn Connor, Abby Doll, Jeffrey A. Drezner, Kristy N. Kamarck, Katherine Pfrommer, Mark V. Arena, Irv Blickstein, William Shelton, and Jerry Sollinger, *Joint Precision Approach and Landing System Nunn-McCurdy Breach Root Cause Analysis and Portfolio Assessment Metrics for DoD Weapons Systems*, Vol. 8, Santa Monica, Calif.: RAND Corporation, MG-1171/8-OSD, 2015. As of March 30, 2017:
https://www.rand.org/pubs/monographs/MG1171z8.html

Killing, J. Peter, "How to Make a Global Joint Venture Work," *Harvard Business Review*, May 1982. As of March 30, 2017:
https://hbr.org/1982/05/how-to-make-a-global-joint-venture-work

Killing, J. Peter, *Strategies for Joint Venture Success*, New York: Routledge, 2014.

Kinzig, Bill, and Dave Bailey, *T-6A Texan II Systems Engineering Case Study*, Wright-Patterson Air Force Base, Ohio: Center for Systems Engineering at the Air Force Institute of Technology, 2010. As of May 9, 2017:
http://www.dtic.mil/dtic/tr/fulltext/u2/a538810.pdf

Kotter, John P., *Leading Change*, Boston: Harvard Business Review Press, 1996. As of May 8, 2017:
https://www.kotterinternational.com/8-steps-process-for-leading-change

Kotter, John P., "Accelerate!" *Harvard Business Review*, November 2012. As of March 30, 2017:
https://hbr.org/2012/11/accelerate

Kotter, John P., *Accelerate: Building Strategic Agility for a Faster Moving World*, Boston: Harvard Business Review Press, 2014.

LaGrone, Sam, "Navy Pondering Helicopter Future After MH-60 Seahawk," *U.S. Naval Institute News*, February 24, 2016. As of March 8, 2016: http://news.usni.org/2016/02/24/navy-pondering-helicopter-future-after-mh-60-seahawk

Lorell, Mark A., Michael Kennedy, Robert S. Leonard, Ken Munson, Shmuel Abramzon, David L. An, and Robert A. Guffey, *Do Joint Fighter Programs Save Money?* Santa Monica, Calif.: RAND Corporation, MG-1225-AF, 2013. As of March 30, 2017: https://www.rand.org/pubs/monographs/MG1225.html

Lorell, Mark A., Robert S. Leonard, and Abby Doll, *Extreme Cost Growth: Themes from Six U.S. Air Force Defense Acquisition Programs*, Santa Monica, Calif.: RAND Corporation, RR-630-AF, 2015. As of May 8, 2017: https://www.rand.org/pubs/research_reports/RR630.html

Lorell, Mark A., Julia F. Lowell, Michael Kennedy, and Hugh P. Levaux, *Cheaper, Faster, Better? Commercial Approaches to Weapons Acquisition*, Santa Monica, Calif.: RAND Corporation, MR-1147-AF, 2000. As of May 9, 2017: https://www.rand.org/pubs/monograph_reports/MR1147.html

Lorell, Mark A., Leslie Adrienne Payne, and Karishma R. Mehta, *Program Characteristics That Contribute to Cost Growth: A Comparison of Air Force Major Defense Acquisition Programs*, Santa Monica, Calif.: RAND Corporation, RR-1761-AF, 2017. As of July 11, 2017: https://www.rand.org/pubs/research_reports/RR1761.html

Marine Aviation, *Marine Aviation Plan 2016*, Washington, D.C.: Headquarters, Marine Corps, Department of Aviation, March 2016.

McKnight, Terry, "Commonality Drives Savings in Shipbuilding," *U.S. Naval Institute News*, February 29, 2016.

Mintzberg, Henry, "Organization Design: Fashion or Fit?" *Harvard Business Review*, January 1981. As of May 10, 2017: https://hbr.org/1981/01/organization-design-fashion-or-fit

Moore, Mark H., *Creating Public Value: Strategic Management in Government*, Cambridge, Mass.: Harvard University Press, 1997.

Moore, Mark H., *Recognizing Public Value*, Cambridge, Mass.: Harvard University Press, 2013.

Moyers, Al, "The Long Road: AFOTEC's Two-Plus Decades of V-22 Involvement," Air Force Operational Test and Evaluation Center, August 2007.

Myers, Dominique, "Acquisition Reform—Inside the Silver Bullet: A Comparative Analysis—JDAM Versus F-22," *Acquisition Review Quarterly*, 2002.

Nobel, Carmen, "Clay Christensen's Milkshake Marketing," *Harvard Business School Working Knowledge Articles*, February 14, 2011. As of March 30, 2017: http://hbswk.hbs.edu/item/clay-christensens-milkshake-marketing

Office of the Assistant to the Secretary of Defense, *Joint Service Chemical and Biological Defense Program FY 06-07 Overview*, Washington, D.C.: U.S. Department of Defense, 2006. As of May 9, 2017: http://www.dtic.mil/dtic/tr/fulltext/u2/a453539.pdf

Office of the Joint Chiefs of Staff, *Department of Defense FY 2013 President's Budget Submission: The Joint Staff*, Washington, D.C., February 2012.

Office of the Joint Chiefs of Staff, *Joint Integrated Air and Missile Defense: Vision 2020*, Washington, D.C., December 5, 2013.

Office of the Joint Chiefs of Staff, *Department of Defense FY 2016 President's Budget Submission: The Joint Staff*, Washington, D.C., January 20, 2015.

Office of the Under Secretary of Defense (Acquisition, Technology, and Logistics), *FVL Capabilities Based Assessment (CBA)*, presentation at AHS International Specialists Meeting, Unmanned Robot Systems, January 21, 2009. As of May 17, 2017: https://vtol.org/store/product/future-vertical-lift-fvl-capabilities-based-assessment-4717.cfm

Office of the Under Secretary of Defense for Acquisition, Technology and Logistics, "Advance Questions for Frank Kendall, Nominee to Be Under Secretary of Defense for Acquisition, Technology, and Logistics," March 2012. As of January 26, 2016: http://www.acq.osd.mil/fo/docs/Kendall%20APQs%20-%2029Mar2012.pdf

OUSD(AT&L)—*See* Office of the Under Secretary of Defense for Acquisition, Technology and Logistics.

Seligman, Lara, "As Teams Vie for JSTARS, Raytheon Stays Above the Fray," *Defense News*, September 12, 2015. As of October 13, 2015: http://www.defensenews.com/story/defense/show-daily/afa/2015/09/12/teams-vie-jstars-raytheon-stays-above-fray/72039724

Simons, Robert, "Stress-Test Your Strategy: The 7 Questions to Ask," *Harvard Business Review*, November 2010. As of October 1, 2015: https://hbr.org/2010/11/stress-test-your-strategy-the-7-questions-to-ask

Smallman, Laurence, Hanlin Tang, John F. Schank, and Stephanie Pezard, *Shared Modular Build of Warships: How a Shared Build Can Support Future Shipbuilding*, Santa Monica, Calif.: RAND Corporation, TR-852-NAVY, 2011. As of May 9, 2017: https://www.rand.org/pubs/technical_reports/TR852.html

U.S. Army, *Initial Capability Document (ICD) for Future Vertical Lift (FVL) Family of Systems (FoS)*, April 8, 2013, not available to the general public.

U.S. Army, *National Commission on the Future of the Army: Aviation Restructure Initiative (ARI)*, May 18, 2015, not available to the general public.

U.S. Army Research, Development and Engineering Command and U.S. Army Aviation and Missile Research Development and Engineering Center, "Joint Multi Role Technology Demonstrator (JMR TD) Update," October 29, 2014. As of May 17, 2017:
http://www.ahs-hrc.org/uploads/1/3/2/7/13275751/jmr_td_brief_for_ahs__29_oct_14_final.pdf

U.S. Department of Defense, *Joint Precision Approach and Landing System, Updated Analysis of Alternatives (AoA) Study Report*, Washington, D.C., November 17, 2005, not available to the general public.

U.S. Department of Defense, *Study on Rotorcraft Survivability: Report to Congress*, Washington, D.C., October 5, 2009.

U.S. Department of Defense, *Joint Direct Attack Munition (JDAM)*, Joint Program Office, Selected Acquisition Report, December 2014.

U.S. Department of Defense, *Joint Light Tactical Vehicle*, Joint Program Office, Selected Acquisition Report, December 2015. As of May 17, 2017:
http://www.dod.mil/pubs/foi/Reading_Room/Selected_Acquisition_Reports/16-F-0402_DOC_43_JLTV_DEC_2015_SAR.pdf

U.S. Department of the Navy, *Naval Aviation Vision 2016–2025*, January 2016. As of May 16, 2017:
http://www.navy.mil/strategic/Naval_Aviation_Vision.pdf

U.S. Government Accountability Office, *Status of the Utility Tactical Transport Aircraft System Program*, Washington, D.C., PSAD-77-31, February 25, 1977.

U.S. Government Accountability Office, *The Light Airborne Multi-Purpose System, LAMPS MK III, Progress Evident but Some Problems and Questions Remain*, Washington, D.C., C-MASAD-81-4, February 23, 1981.

U.S. Government Accountability Office, *Joint Major System Acquisition by the Military Services: An Elusive Strategy*, Washington, D.C., NSIAD-84-22, December 1983.

U.S. Government Accountability Office, *The V-22 Osprey—Progress and Problems*, Washington, D.C., NSIAD-91-45, 1990. As of May 17, 2017:
http://www.gao.gov/products/NSIAD-91-45

U.S. Government Accountability Office, *Precision Guided Munitions in Inventory, Production and Development*, Washington, D.C., GAO/NSAID-95-95, June 1995.

U.S. Government Accountability Office, *Acquisition Plans for Training Aircraft Should Be Reevaluated*, Washington, D.C., GAO/NSAID-97-172, September 1997.

U.S. Government Accountability Office, *Challenges Associated with Implementing the Joint Tactical Radio System*, Washington, D.C., GAO/NSAID-99-179, September 1999. As of May 8, 2017:
http://www.gao.gov/archive/1999/ns99179.pdf

U.S. Government Accountability Office, *Restructured JTRS Program Reduces Risk, but Significant Challenges Remain*, GAO-06-955, September 2006. As of May 10, 2017:
http://www.gao.gov/assets/260/251421.pdf

U.S. Government Accountability Office, *Defense Acquisition Process: Military Service Chiefs' Concerns Reflect Need to Better Define Requirements Before Programs Start*, Washington, D.C., GAO-15-469, June 11, 2015.

Vertical Lift Consortium, "The Vertical Lift Consortium Views and Recommendations for the Future Vertical Lift Strategic Plan," unpublished white paper, 2015a.

Vertical Lift Consortium, presentation briefing to the Army Science Board, July 24, 2015b.

VLC—*See* Vertical Lift Consortium.

Weisgerber, Marcus, "Should One U.S. Service Rule the Military's Drones?" *Defense One*, August 24, 2015. As of March 30, 2017:
http://www.defenseone.com/management/2015/08/should-one-service-rule-military-drones/119399

Weisgerber, Marcus, "F-35 Chief: Think Very, Very Hard Before Making Another Joint Fighter," *Defense One*, March 11, 2016. As of March 30, 2017:
http://www.defenseone.com/technology/2016/03/f-35-chief-think-very-very-hard-making-another-joint-fighter/126587/?oref=d-river

Weiss, Geoffrey, "Seeing 2020: America's New Vision for Integrated Air and Missile Defense," *Joint Force Quarterly 76*, National Defense University Press, December 30, 2014. As of May 9, 2017:
http://ndupress.ndu.edu/Media/News/News-Article-View/Article/577599/jfq-76-seeing-2020-americas-new-vision-for-integrated-air-and-missile-defense

Woolsey, James P., Harold S. Balaban, Kristen M. Guerrera, and Bruce R. Harmon, *Analysis of the JSF Engine Competition*, Alexandria, Va.: Institute for Defense Analyses, IDA Document NS D-4661, July 2012.

Wunker, Stephen, "Six Steps to Put Christensen's Jobs-to-Be-Done Theory into Practice," *Forbes*, February 7, 2012.

Yan, Aimin, and Yadong Luo, *International Joint Ventures: Theory and Practice*, Armonk, N.Y.: M.E. Sharpe, 2001.